# Text Mining with Machine Learning
## Principles and Techniques

**Jan Žižka**
Machine Learning Consultant, Brno, Czech Republic

**František Dařena**
Department of Informatics
Mendel University, Brno, Czech Republic

**Arnošt Svoboda**
Department of Applied Mathematics & Computer Science
Masaryk University, Brno, Czech Republic

CRC Press
Taylor & Francis Group
Boca Raton  London  New York

CRC Press is an imprint of the
Taylor & Francis Group, an **informa** business

A SCIENCE PUBLISHERS BOOK

CRC Press
Taylor & Francis Group
6000 Broken Sound Parkway NW, Suite 300
Boca Raton, FL 33487-2742

Version Date: 20190826

International Standard Book Number-13: 978-1-138-60182-6 (Hardback)

Library of Congress Cataloging-in-Publication Data

Names: Žižka, Jan, author. | Dařena, František, 1979- author. |
  Svoboda, Arnošt, 1949- author.
Title: Text mining with machine learning : principles and techniques / Jan
  Žižka, František Dařena, Arnošt Svoboda.
Description: First. | Boca Raton : CRC Press, 2019. | Includes
  bibliographical references and index. | Summary: "This book provides a
  perspective on the application of machine learning-based methods in
  knowledge discovery from natural languages texts. By analysing various
  data sets, conclusions, which are not normally evident, emerge and can
  be used for various purposes and applications. The book provides
  explanations of principles of time-proven machine learning algorithms
  applied in text mining together with step-by-step demonstrations of how
  to reveal the semantic contents in real-world datasets using the popular
  R-language with its implemented machine learning algorithms. The book is
  not only aimed at IT specialists, but is meant for a wider audience that
  needs to process big sets of text documents and has basic knowledge of
  the subject, e.g. e-mail service providers, online shoppers, librarians,
  etc"-- Provided by publisher.
Identifiers: LCCN 2019035868 | ISBN 9781138601826 (hardback)
Subjects: LCSH: Machine learning. | Computational linguistics. |
  Semantics--Data processing.
Classification: LCC Q325.5 .Z59 2019 | DDC 006.3/12--dc23
LC record available at https://lccn.loc.gov/2019035868

Visit the Taylor & Francis Web site at
http://www.taylorandfrancis.com

and the CRC Press Web site at
http://www.crcpress.com

*To our families*

# Preface

The aim of this book is to make it easier for those interested in text mining – one of the branches of modern information technology – to enter an interesting area focused on analyzing texts in natural languages. Machine learning has been chosen as an analytical tool because it is a promising organic part of computer science and artificial intelligence that supports successful exploration of big volumes of text data using advanced classification and clustering methods. Machine learning now has available a large number of software tools, both commercial and free, so there is no need to implement them and a researcher can focus directly on their use for given data and applications.

The authors of this book have chosen demonstrations of certain basic as well as popular implementations of machine learning algorithms in an advanced interpreted programming language R, which is freely available for a variety of operating systems and provides an appropriate communication interface over RStudio without the need for complex learning. In no case is this book primarily focused on advanced computer scientists, programmers, or mathematicians, although basic programming skills and higher mathematical knowledge (linear algebra, statistics, function analysis) are required to understand algorithms, use them and, of course, interpret their results. For interested readers, chapters have references to literature which can suitably complement the topics described. Text mining has already come to an advanced/a well-developed form, driven by the need to explore texts as specific data type, in particular large text data files on the Internet from social networks, product or service reviews, e-mails, blogs, and many forums. Perhaps every e-mail user today is aware of the facility to filter messages (for example, spam removal or topic grouping), which is just one of many possible applications of the text mining industry.

The book covers the introduction to text mining by machine learning, introduction to the R programming language, structured text representation, classification and easy introduction to the most common classification algorithms (Bayes classifier, nearest neighbors, decision trees, random forest, support vector machines, and deep learning), followed by clustering. The final two chapters deal with word embedding and feature selection problem areas.

The authors of this book would like to express their hope that the book will find a number of interested readers and will be used to familiarize them with the attractive sector of artificial intelligence and machine learning applied to the difficult area of processing data produced by humans and not by machines.

# Contents

# Authors' Biographies

**Jan Žižka** is currently involved in consultations in the field of machine learning and data mining. For many years he worked as a system programmer, developer of advanced software systems, and researcher. For the last 25 years, he has devoted himself to artificial intelligence and machine learning, especially text mining. He has also worked at a number of universities in the Czech Republic and abroad, including research institutes. He is the author and co-author of approximately 100 international publications.

**František Dařena** works as an associate professor, the guarantor of the doctoral degree program System Engineering and Informatics, and head of the Text Mining and Natural Language Processing group at the Department of Informatics, Faculty of Business and Economics, Mendel University Brno. He is the author of several publications in international scientific journals, conference proceedings, and monographs; member of the editorial boards of international journals, and editor-in-chief of International Journal in Foundations of Computer Science & Technology. His research areas include text/data mining, intelligent data processing, and machine learning.

**Arnošt Svoboda** is an expert in programming and his area includes programming languages and systems such as R, Assembler, Matlab, PL/1, Cobol, Fortran, Pascal, and others. He started his career as a system programmer. For the last 20 years, Arnošt has worked as a teacher and researcher at Masaryk University in Brno (Czech Republic), Faculty of Economics and Administration, Department of Applied Mathematics and Computer Science. At present, he is interested in machine learning and data mining.

# Chapter 1

# Introduction to Text Mining with Machine Learning

## 1.1 Introduction

In the last few years, we have seen an explosion in the amount of data as the result of many activities. The data comes from various sources and is available in many formats. This has been enabled, primarily, by the massive advance on the Internet. Various devices like smart phones enable communication and the use of applications anytime and anywhere; many activities like shopping, interacting with government institutions, or providing support to customers, are moving to digital environments; many documents are being digitalized; people meet and interact on virtual platforms. The transformation of the Web into Web 2.0 [26], where the content is actively created by users, has thrown open to the masses, many possible avenues to express their ideas, recommendations, or attitudes.

It is obvious that texts written in a natural language are a natural way of human communication. Textual documents are thus strongly related to many human activities and they have become a source that is worth analyzing. The results of such an analysis can bring useful insights in practically all domains.

As the ability of people to analyze texts remains essentially the same, the availability of more data requires new computational methods in order to find something useful in large collections of documents. Thus, a discipline known as *text mining* has become very popular and attractive. Text mining can be defined as a knowledge-intensive process in which a user interacts with a collection of documents by using analytic tools in order to identify and explore interesting patterns [87]. Applications can be found in marketing, competitive intelligence, banking, health care, manufacturing, security, natural sciences, and many other domains [254, 192].

Computers are able to analyze only the *syntactic* aspect of texts, which means that they are able to recognize how words are arranged in the documents. Because texts in a natural language are written using a grammar, some syntactic patterns in a text can be more or less easily identified. *Semantics* refers to the meaning of a word or group of words in a context. Without perfect understanding of a language it is not possible to completely understand the meaning. Fortunately, is is possible to solve many practical problems even without the full understanding of a text because syntax and semantics are often closely related. If two texts use the same words and syntactic structures, they are likely to be semantically similar and can, for example, belong to the same class of documents [196].

Documents can generally be analyzed in two different ways. The statistical or machine learning approach uses mathematical representation of texts. Linguistic methods, using natural language processing techniques, represent texts using language models where the meaning and different relationships are contained. Text mining uses both approaches to find knowledge, usually in a large number of texts [254, 271].

## 1.2   Relation of Text Mining to Data Mining

Text mining encompasses a wide variety of tasks that can bring information about different aspects of texts. The typical text mining tasks include [196, 279, 123]:

- categorization of documents – assigning a document to one or more pre-defined category (e.g., assigning a newspaper article to one or more categories, labeling e-mails as spam or ham);

- clustering – grouping documents according to their similarity, for example, in order to identify documents sharing a common topic;

- summarization – finding the most important parts in one or more documents and creating a text that is significantly shorter than the original;

- information retrieval – retrieving documents that match a query representing information needed from a large collection of documents;

- extracting the meaning of documents or their parts – identifying hidden topics, analyzing sentiment, opinion, or emotions;

- information extraction – extracting structured information like entities, events, or relations from unstructured texts;

■ association mining – finding associations between concepts or terms in texts;

■ trend analysis – looking at how concepts contained in documents change in time;

■ machine translation – converting a text written in one language to a text in another language.

Some of the text mining tasks are very similar to the tasks of *data mining*. Data mining is the automatic or semiautomatic process of finding implicit, previously unknown, and potentially useful knowledge in collections of electronically stored data. The knowledge has a form of structural patterns in data that can be also used to make predictions or provide answers in the future [280].

Data mining includes many different methods, tools, algorithms, or models. All of them require the data to be in a structured form. This means that the data can be represented in a tabular form as in a relational database. The data takes the form of a set of examples (or instances, data points, observations) described by specific values of their features (or attributes, variables, fields).

The features can be of several types [75, 174]:

■ categorical (nominal) – the domain is a discrete set of values where ordering does not make sense;

■ binary – a special type of categorical attribute with only two possible values;

■ ordinal – the domain is a discrete set of values that can be ordered;

■ numerical – the value of a feature is a number, either integer or continuous.

An example of a structured representation of data describing purchases in a retail store is in Table 1.1. Every purchase is characterized by a customer (described by age, education level, and gender), date of purchase, and the total price.

Such a format is, however, not typical for texts. Generally, a text is a string written in a natural language consisting of parts (words) with certain meaning that are combined according to some rules (syntax). The texts can also be of a

**Table 1.1:** A structured representation of purchases in a retail store.

| Age | Gender | Education | Date | Price |
|-----|--------|-----------|------|-------|
| 35 | female | primary | 2019-02-10 | 20.0 |
| 40 | male | tertiary | 2019-02-14 | 28.4 |
| 21 | male | secondary | 2019-01-30 | 15.1 |
| 63 | female | secondary | 2019-03-01 | 11.9 |

different range. A unit of a text can be a sentence, a few sentences combined in a paragraph, or much longer texts, like web pages, e-mails, articles, or books. Sometimes, a text can be just a few words that are not a valid sentence, which is quite typical, for example, for short posts on social networks.

In order to be able to apply data mining methods to texts, they need to be converted to a structured representation. A classic structured representation of texts as vectors in a *vector space model* is known as *bag-of-words* and the process of inferring this representation is described in chapter 3. Another, more modern, representation that is based on embedding words to a continuous vectors space referred as to *word embeddings* is discussed in Chapter 13.

There are a few problems related to the bag-of-words representation, which are not that typical for data mining tasks in general [134, 65, 188, 170, 202]:

■ Independently of the complexity of features, the *input space* for text mining problems *is large*. It is not uncommon that the dictionaries of various natural languages contain hundreds of thousands of words. When considering not only words as features but also, for example, combinations, mutual positions, or grammatical relationships between words, the complexity further increases. Of course, not all words of a language will appear in most document collections. On the other hand, even relatively small collections can contain tens of thousands of unique words (see Figure 1.1).

■ The data contains a lot of *noise* which is often given by the nature of the data. Text documents are usually created by people that make many errors, including mistyping, grammatical errors, and incorrect labeling (e.g., assigning one star instead of five stars to a positive review).

■ The vectors representing documents are very *sparse*. If we compare the number of unique words in one document (tens of words for short messages) to the dictionary size (tens of thousands), the difference is enormous. Thus, only a small fraction of vector elements are non-zero values (see Figure 1.1).

■ The *distribution* of the probability with which the words appear in documents is strongly *skewed*, following so called Zipf's law (see Section 3.3). This means that the majority of the content is represented with just a few words and that most of the words appear with very small frequencies.

■ In many tasks, only a small fraction of the content is often relevant. For example, for a document classification task, only a few hundred words from the dictionary are needed in order to correctly assign documents to their classes with a relatively high accuracy. Many of the words are therefore *redundant*.

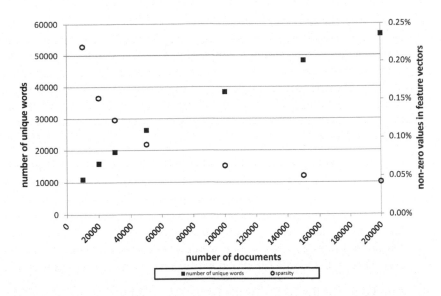

**Figure 1.1:** The graph represents a dependency between the number of documents and the number of unique words in a document collection. The documents represent customer reviews with the minimal, maximal, and average lengths equal to 1, 400, and 24 words.

■ For many tasks, there often exists a *scarcity of labeled data*. This is because the labeling process, usually performed manually, is very demanding. A special situation is that there are instances of positive training data for a specific problem, but negative examples are missing.

■ In many situations, the distribution of the data in terms of classes or topics is *not balanced*. This means, that some classes have many more instances than the others which is generally a problem for learning. An example is the field of sentiment analysis, where the positive reviews in certain product categories are much more common than the negative ones.

## 1.3   The Text Mining Process

To find useful knowledge in a collection of text documents involves many different steps. To arrange them into a meaningful order, we might look at the general text mining process. It consists of the following steps [271, 133, 196]:

■ **Defining the problem:** This step is actually independent of any actions which may subsequently be taken. Here, the problem domain needs to be understood and the questions to be answered, defined.

- **Collecting the necessary data:** The sources of texts containing the desired information need to be identified and the documents collected. The texts can come from within a company (internal database or archive) or from external sources, from the web, for example. In this case, web scrapers need to be frequently implemented to directly grab the content of the web pages. Alternatively, the API of some web based systems can be used to retrieve the data. After retrieval, texts are stored so they are ready for further analysis.

- **Defining features:** Features that well characterize the texts and are suitable for the given task need to be defined. The features are typically based on the content of the documents. A very simple approach, bag-of-words with binary attribute weighting, takes every word as a boolean feature. Its value indicates whether or not the word is in a document. Other methods might use more complicated weighting schemes or features that are derived from the words (modified words, combinations of words, etc.).

- **Analyzing the data:** This is the process of finding patterns in data. According to the type of task to be solved (e.g., classification), a specific model or algorithm is selected and its properties and parameters defined. Then, the model can be applied to the data so a solution to the solved problem can be found. To solve a specific problem, more models are usually available. The choice is not explicitly given in advance. The models have different characteristics that influence the data mining process and its result. The model can be (white box) or does not have to be (black box) well interpretable by a human. Some models have higher computational complexity than the others. According to the utilization of the model, fast creation can be preferred over fast application or vice versa. The suitability of a model is often strongly dependent on the data. The same model can provide excellent results for one data set while it can completely fail for another. Thus, selecting a proper model, finding the right structure for it, and tuning the parameters often requires a lot of experimental effort.

- **Interpreting the results:** Here, some results are obtained from the analysis. We need to carefully look at them and relate them to the problem we wanted to solve. This phase might include verification and validation steps in order to increase the reliability of the results.

## 1.4 Machine Learning for Text Mining

*Machine learning* [199] as part of computer science and informatics is one of the most practically oriented areas of artificial intelligence [161]. It was inspired by the ability to learn – in other words, to gain new or additional knowledge, which is one of important characteristics of living organisms. Machine learning as a science focuses on looking for and developing algorithms, which can simulate or emulate mental abilities of living organisms from the learning point of view.

Learned knowledge may not completely address everything – for it is something of a general nature – but, it allows us to solve future problems that did not occur frequency, in the same form and/or environment in the past. If such a future problem is similar to something which has already occured, it is often not possible to solve it using the exact instructions from the past; however, a similar solution (or procedure) can hopefully be successful in most (or ideally, all) cases that are yet to occur.

Machine learning is one of the modern tools that includes anything which can be useful. Today, it consists of tens of various algorithms, and research constantly brings modified or completely new algorithms, sometimes replacing the older ones because the field of machine learning is strongly dominated by the practical needs of the real world that are growing in time. In addition to algorithms and relevant information technologies, machine learning inevitably uses mathematics, in particular the theory of probability, statistics, combinatorics, mathematical analysis, and a set of other essential disciplines.

To solve problems with help of computers, machine learning differs from the traditional method, which depends on applying carefully implemented and debugged programs to input data. The traditional computer program consists of sets of instructions that process the input data and provide the output using a set of predefined parameters, which are provided externally. An example is a computer program that searches for a pre-set list of solutions to expected problems, something like a table containing results for combinations of input values or a set of mathematical equations. Such a program can be–in a simplified way– considered as a function $f(x) = y$, which for a given input $x$ returns an output $y$, however, provided that $f(x)$ is reliably known. Without knowing such a function, it is necessary to estimate or approximate it.

If the needed conditions and rules are observed, the result is guaranteed because it is based on mathematically proved theorems and methods. Unfortunately, if there are–often even very small– changes in data or working environment, such a program is predisposed to a fundamental error or malfunction. For example, analyzed text data may begin to contain new terms that were not known when the program was created, or it becomes necessary to start analyzing such data in another language. In such a case, it is inevitable to rework, often completely, that program.

Machine learning can use an existing implemented algorithm, retrain it with other examples and apply such new or extended knowledge to solving a new problem. A certain defect might be that this knowledge is not supported enough by mathematical proofs – in machine learning, the support comes from empirical proofs and heuristics.

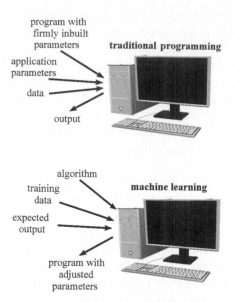

**Figure 1.2:** Traditional programming versus machine learning – illustration of the principles.

A simple illustration of the difference between traditional methods and machine learning is shown in Figure 1.2.

Determining the structure of models and their parameters is the goal of the *training* process, which uses (often iteratively) the presented training samples for this purpose. The success rate of the training phase must be measured by *testing*, which uses another set of samples which are not part of the training process.

## 1.4.1 Inductive Machine Learning

Current methods, oriented on solving practical problems for example [221, 208, 75], are particularly based on *inductive* approaches. The inductive learning is based on the generalization of (many) specific samples for a given problem area. The samples are comprised of collected *data*. For purpose of this book, that part of data, which is relevant for solving a given problem, is called *information*. Then, generalized information forms *knowledge*. The main task of machine learning is mining knowledge hidden in large collections of data.

Having a set of $n$ samples of data items, $\{x_1, x_2, \ldots, x_n\}$, and knowing for each sample $x_i$ its correct result $y_i$, the approximate function $\hat{f}(x) \approx f(x)$. The accuracy of such approximation depends naturally on the quality of samples – how well they can cover (or represent) all possible data items. In the real world, typically, not all possible data items are available for a given moment, or there might be so high (whether final or infinite) a number of such items that it is practically impossible to process them within an available time period. At the very least, data that may occur in the future may not be known now and moreover, may not be fully represented by familiar data from the past.

Unlike the traditional applications of computer programs, machine learning aims at automatically discovering the appropriate parameters of a selected algorithm, the implementation of which may already be available, but which needs its parameters adjusted to specific data.

## 1.5 Three Fundamental Learning Directions

Depending on information provided by collected teaching samples, the current machine learning sphere distinguishes three fundamental directions:

■ Supervised learning

■ Unsupervised learning

■ Semi-supervised learning

### 1.5.1 Supervised Machine Learning

Known also as learning with a teacher, supervised machine learning relies on labeled training samples. Each sample $\mathbf{x}_i$ has its corresponding function value $y_i$:

$$y_i = f(\mathbf{x}_i) .$$

If the function $y_i$ is defined as non-numerical, its values are labels of classes. For example, books may be classified according to verbal evaluation by their readers who also assign a decision (label) $y_i \in \{interesting, tedious, neutral\}$. A classification algorithm learns using the labeled samples. This learning process is called *learning with a teacher* because the trained algorithm has feedback based on the knowledge of the correct answers (labels). The feedback is provided by the 'teacher' that supervises the correctness of classifications and the 'learner' knows whether its answer was right or not so that it can improve its knowledge – appropriately adjusted parameters of the algorithm.

The regression methods for numeric functions are not discussed because they are not relevant here.

### 1.5.2 Unsupervised Machine Learning

Unlike supervised machine learning, the unsupervised version misses feedback from the absent teacher, thus the learners must learn themselves. The term 'absent teacher' means that the available training samples $\mathbf{x}_i$ lack their appropriate class labels, therefore it is not directly possible to reveal what is relevant for each class. In addition, it is not known what (and how many) classes exist for a given case. The task is now not only to learn how to classify correctly but also to which classes.

This branch of machine learning is called *learning without a teacher* and employs various clustering algorithms and methods. Based on the contents of $x_i$'s, the clustering algorithms try to detect which samples are similar enough to establish their particular cluster. Then, each cluster represents a class. Usually, it is necessary to define (classes) in advance how many clusters should be generated; this is a frequent problem. Too few clusters can mean too vague knowledge mined from the data. The more clusters, the more specific each of them is; however, too many clusters lack the property of being generalized knowledge extracted from the samples – in the extreme case, each cluster contains only one case, which is not very useful knowledge.

There exist also certain clustering algorithms that are able to discover the number of convenient clusters; for example, the EM (expectation-maximization) algorithm is worth mentioning. For more details, see [112] chapter 8.5. EM assigns a probability distribution to each instance which indicates the probability of its belonging to each of the clusters. EM can decide how many clusters to create by cross validation, or it can be specified *a priori* how many clusters to generate.

The unsupervised learning belongs to more difficult and demanding learning methods because it must work with less information. However, clustering today constitutes one of the highly needed applications in the real world. An interested reader can find more information concerning the clustering methods in [142, 230, 22, 254], and many others.

### 1.5.3 Semi-supervised Machine Learning

A modern machine learning branch is *semi-supervised machine learning*, which – as its name suggests – represents something in between the extreme cases of supervised and unsupervised learning methods. In semi-supervised machine learning, the learner is given only a few labeled training samples and many unlabeled ones.

The labeled samples provide some valuable initial information, which makes it easier to separate (and label) unlabeled training samples so that the correct classes can be generated.

This task is typical for contemporary situations when it is relatively much easier to automatically collect and store a lot of unlabeled data than to laboriously label them all.

As a result of the use of modern information and communication technologies (the Internet, social networks), while there is an increasing amount of generally valuable (and not only) textual data on a daily basis their processing, for example from a semantic point of view, is now completely beyond human 'manual' processing and research.

A limited number of training samples gives a limited amount of information which can, however, be used advantageously for advanced text analysis. In any case, this limited information provides some guidance on how to deal with incomplete (unlabeled) data items. A teacher can prepare a small set of samples with answers (classification results), which may be used by the learner as a kind of clue to help solve the class assignment with respect to non-classified training items. At the same time, it makes it possible to better determine the number and types of suitable classification classes.

A good introduction with examples of several algorithms (Self-training, Co-training, and some others) can be found in [1].

## 1.6 Big Data

One of the problems that we can experience in text mining is that there are too many data examples available. On the one hand, it can be considered positive in terms of sufficient coverage of data with training examples. However, on the other hand, it usually leads to great difficulties due to too high computational complexity (time and memory), which the available hardware may not be able to handle. This is a problem known as *Big Data* (which is, however, not discussed in this book; as an introduction, a reader may refer to, for example, [147, 152]).

For text mining, the mentioned big-data problem is often characteristic. The reason is that, thanks to broadly available world-wide web technologies like the Internet, textual data accumulate incredibly fast every day. Various social networks belong to most generous data providers and in principle, they represent a very valuable source of information not only for the contributors. Unfortunately, it is not known how to define a mathematical function, which would be able to process that natural-language based textual data so that the data may be properly analyzed from its semantic point of view like human researchers can. Because of the huge amount of the data, it is necessary to use informatics and computers, which is where machine learning helps.

## 1.7 About This Book

Today, a lot of good publications and books are already available for familiarization with machine learning, for example [199, 151, 187, 261]. Tens of books from various publishers deal with many different problem areas from the machine learning point of view, and a reader will, undoubtedly be able to find what her or his specific interests need. Our goal is, therefore, not to teach machine learning in general. Instead, we would like to show how machine learning can be used to solve practical tasks in a specific domain, which is text data analysis.

In order to obtain some real results and enable readers to implement some of the described procedures, we decided to use the R programming language. R is a great statistical software package with many advantages. It is free, it is interpreted (the same code can be run on many platforms), and provides an extensive functionality contained in many available packages. It is used by many researchers and practitioners all over the world. The solutions of problems using R are, however, often not suitable for large scale problems because R works only with data that fit to the memory of a computer and is often slow (the reduced speed is also because it is very flexible through dynamic typing).

In R and its available libraries, many machine learning algorithms are already implemented. It is, therefore, necessary to provide them with prepared data in a desired format, get the results, and display, visualize, or evaluate them. All these steps are combined in computer programs (scripts) that are run on an R interpreter.

In individual chapters dealing with typical text mining tasks, the code that is used to demonstrate the algorithms and solve some practical problems is also presented. To enable readers to understand the code, some of the R basics are included in the book too. We do not try to explain all aspects of the language, all available algorithm implementations, their possible parameters, inputs, and outputs, but only what is necessary for our specific examples. All other details can always be found in the newest description (manual) of the used R package.

# Chapter 2

# Introduction to R

R is a high-level programming language and environment for statistical computing and graphics. For many years, R has belonged to the most popular programming languages, especially in the machine learning community. It is primarily focused on statistical operations, working with vectors and matrices, creating statistical models on a few lines, etc. [209].

R provides facilities for effective data handling and storage, operators for calculations on multidimensional data, a collection of tools for data analysis and graphical facilities for data analysis and display. It also includes a simple and effective programming language which includes all common elements of programming languages, such as branchings, loops, user-defined functions, or input and output facilities [117]. The capabilities of R language can be augmented with compiled code written in C, C++, or Fortran [80].

The language, created by Ross Ihaka and Robert Gentleman at the University of Auckland more than twenty years ago, has derived its syntax, semantics, and implementation from two other older languages – S and Scheme [125]. R is an open-source project, which means that you can see how the functionality is implemented and modify what does not work exactly according to your requirements [203]. R is very popular in science, engineering, and education, with many active communities. The R Foundation supports the R project and its development, organizes conferences, and patronizes the R Journal (*https://journal.r-project.org*).

This book does not want to provide exhaustive information on programming in R. The language is used here as one of many options with the aim of demonstrating the principles of machine learning algorithms and getting real results from analyzing texts. An interested reader can refer to an endless resource of books (e.g., *The R Primer* by C. T. Ekstrom {CRC Press}, *R in a Nutshell: A*

*Desktop Quick Reference* by J. Adler {O'Reilly}, *The R Book* by M. J. Crawley {Wiley}, or *The Art of R Programming: A Tour of Statistical Software Design* by N. Matloff {No Starch Press}. Probably the most compete information can be found in *The R Manuals* available at *https://cran.r-project.org/manuals.html*.

Some of the tasks can be achieved with core R functions and capabilities, some other algorithms are implemented by various packages available in the *Comprehensive R Archive Network* (CRAN). The archive currently contains almost 14,000 packages solving problems from various domains and uploaded by many authors. To solve one specific task, multiple packages are often available. In this book, we do not consider all possible options as the main focus is not on R but the machine learning algorithms. Also only the necessary functionality (only some functions and their parameters) of the packages is used, which is sometimes only a small fraction of the package capabilities. The complete functionality and parametrization of algorithms can be always found in the package documentation.

## 2.1 Installing R

We assume that you have a PC or an Apple Mac and you want to install R on the hard disc and that you have access to the Internet. The R language is available as free software under the terms of the Free Software Foundation's GNU General Public License (*https://www.gnu.org*) in source code form at *https://www.r-project.org*. It can be compiled and run on a wide variety of UNIX platforms, as well as on Windows and Mac OS X.

In order to get a current version of the language, go to the CRAN site (*https://cran.r-project.org/mirrors.html*) and choose your preferred CRAN mirror. Then, you can decide the platform (Linux, Max OS X, or Windows) for which to obtain a precompiled binary distribution. You can also get the source code for all platforms (see Figure 2.1).

In our example, a version for the Windows operating system will be installed. The installation process for all other supported systems is described on the website of the R project.

There exist a few possibilities of the R download for Windows. The base sub-directory is preferred for the first installation.

While older distributions are still accessible, the newest distribution is recommended for downloading. The executable distribution file (e.g., *R-3.5.1-win.exe*) can be easily downloaded and run. R will be installed in an ordinary process with the assistance of an installation wizard.

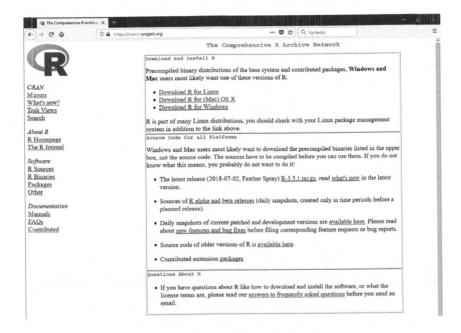

**Figure 2.1:** The home page of the R project.

## 2.2 Running R

To run R under Windows, just click on the R icon. A graphical user interface, as in Figure 2.2, will be displayed.

Inside the R graphical user interface, you can see a menu, toolbar, and R Console. In the console, below the header (a version of R and some initial information about the system) you will see a line with a > symbol on the left. This is called a prompt and this is where you type your commands. Try to enter citation() to see how to cite the R software in your work.

An easier way to use R is by using some of the integrated development environments (IDE). A list of some R IDEs includes:

■ R Studio,

■ Visual Studio for R,

■ Eclipse,

■ Red-R,

■ Rattle,

■ and many others.

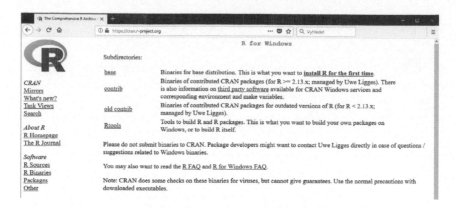

**Figure 2.2:** Selecting a subdirectory for R installation.

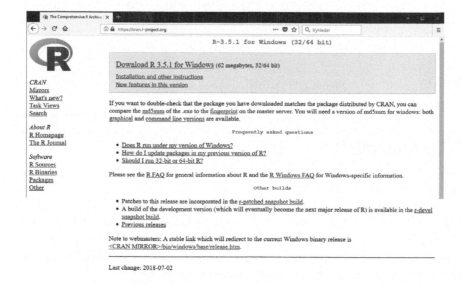

**Figure 2.3:** Getting the newest R package.

*RStudio* has been voted the most popular R IDE by the readers of KD-nuggets[1]. Thus, in the following text, we are using RStudio for the examples.

---

[1] https://www.kdnuggets.com/polls/2011/r-gui-used.html

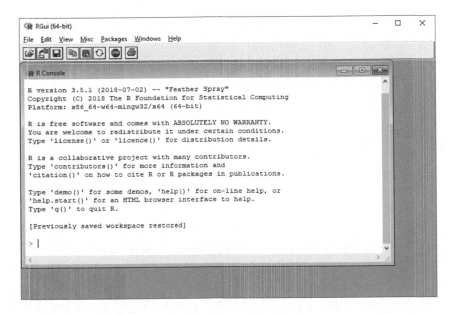

**Figure 2.4:** The graphical user interface of the R system on Windows.

## 2.3 RStudio

RStudio is an IDE for R available in open source and commercial editions. It enables syntax highlighting, code completion, and smart indentation when writing the code that can be executed directly from the source editor. It facilitates the work of programmers by providing facilities like jumping to function definitions, integrated documentation and help, managing multiple working directories, multiple-file editing, interactive debugger for diagnosing and fixing errors, and package development tools.

RStudio can be run on various operating systems like Windows, Mac OS X, and Linux. It is a very useful and powerful tool for programming in R. We suggest that readers use RStudio when learning from this book. To get RStudio, go to *https://www.rstudio.com/* and follow the instructions.

On launching RStudio, the following panes are displayed (see Figure 2.5):

■ *Console* where R commands/scripts can be executed and *Terminal* allowing the access to the system shell (top left-hand corner of the screen)

■ *Environment* and *History*, and *Connections* – information about environment, history of commands, and connections to external data sources (top right-hand corner of the screen)

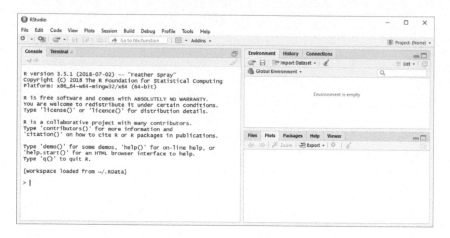

**Figure 2.5:** The graphical user interface of RStudio.

■ *Files* in the current working directory, installed R *packages*, *plots*, a *help* display screen, and a *viewer* of local web content (bottom right of the screen).

A new script, i.e., a code that can be saved on a disk and later reopened or executed, can be created by choosing *File / New File / RScript* in the main menu. A new empty editor window, ready for editing, will appear.

It is always good practice to choose a working directory when programming in RStudio. The working directory is a folder where you store your R code, figures, etc. To set the working directory, go to *Session / Set Working Directory / Choose Directory* in the main menu.

## 2.3.1 Projects

RStudio provides a very useful feature known as *Projects* to organise your work into meaningful contexts. A project puts together R scripts, input data, analytical results, or plots, and has its own working directory, history, or workspace.

A new project can be created by selection of *File / New Project* in the main menu. A project can be started in a new directory, associated with an existing working directory, or cloned from a version control (Git or Subversion) repository.

Select the *New Directory* option, then *New Project*, choose a name for your project directory, and select its location. A new directory will be created, together with a project file (with the *.Rproj* extension) and a hidden directory *.Rproj.user* with project-specific temporary files project. The will be loaded in RStudio and its icon displayed to the right of the toolbar. RStudio will also automatically set the working directory to the project directory.

**Figure 2.6:** A new project created in RStudio.

## 2.3.2  Getting Help

Help pages for a desired object (e.g., operator or function) are invoked by the help() function. The name of an object for which help should be displayed is passed as a string parameter (enclosed in single or double quotes). However, a shortcut ? is often used (quotation marks not necessary). For example, typing ?sin opens a help page containing a documentation for the sin() function.

If you do not remember the precise name of the object but know the domain to which it is related (e.g., trigonometric functions in this case) use the help.search() function. A string passed to the function will be searched for in the documentation and the possibilities associated with this query will be displayed. The same can be achieved with the ?? shorthand (e.g., ??trigonometric or ??"trigonometric functions").

Another way to get help in RStudio is to select *Help / R Help* in the main menu or by going to the Help pane and browsing the manual pages or searching for a specific answer using the search capability.

To see some examples related to, for instance, a function, use the example() function, where the appropriate name is passed as a parameter (e.g., example(help) or example('+')).

## 2.4  Writing and Executing Commands

When R is ready to accept commands (expressions, consisting of objects and functions) from a user, the R console displays a > prompt. After it receives a

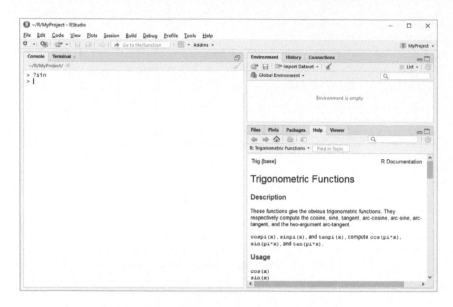

**Figure 2.7:** Displaying a help for an existing function.

command (something is typed on the keyboard and *Enter* is pressed) R tries to execute it. Subsequently, a result is shown[2] in case of a successful execution and a new > prompt is displayed again.

```
> 3+8/4  # after writing the expression, Enter was pressed
[1] 5
> "Hello, World!"
[1] "Hello, World!"
>
```

When the command is not complete (for example, a closing parenthesis, quote, or operand is missing) R will submit a request to finish it. The console will now display a + prompt. The user can complete the command and the result will then be displayed. To finish writing the command when the + prompt is displayed, just press *Esc*.

---

[2]The number in square brackets is related to numbering the elements of the output and can be ignored. Here, for example, the output is a vector and the row starts with its first element.

```
> 5+      # Enter was pressed here
+ 6       # Enter was pressed here
[1] 11
> 5+      # Enter was pressed here
+         # Esc was pressed here

>
```

Two or more expressions can be placed on a single line. In that case they must be separated by semicolons with optional white spaces:

```
> 1+2;2-3; 4/5
[1] 3
[1] -1
[1] 0.8
```

To complete a piece of code, press *Tab*. A list of possible completion options (together with their short descriptions when applicable) that fit the given part of code is displayed (RStudio analyzes the code to suggest only meaningful options). By selecting one of them, the selected piece of text will be inserted to a command. If there exists only one possibility of what can be inserted no options are displayed and the code is completed immediately.

By pressing the up and down arrow keys, previously entered commands can be accessed. Pressing *Ctrl+Up* displays the list of all commands. They are also accessible in the History pane of RStudio.

Comments are useful pieces of the code that help programmers explain what various part of the code mean. They are completely ignored by the R interpreter. A comment starts with the symbol #. The subsequent characters until the end of line are the comment. R does not allow comments running into multiple lines. To overcome this problem, a quoted string occupying more than one line can be used. The string will be evaluated as normal but will not perform any action.

## 2.5   Variables and Data Types

To store and access values in computer memory, so called 'objects' are used. Objects are data structures containing values of a certain type. Every such a structure might have a class attribute, which determines the class of the object, associated to it. If such an attribute does not exist, it will be computed from the type of the values stored in the object.

There exist a few atomic (single-valued) data types (classes, returned by the `class()` function as mentioned below) that are typically used in machine learning:

- character,

- numeric,

- integer,

- logical.

The basic data structure in R, where the atomic values are combined, is a vector; there are no scalars. The other structures include matrices, arrays, factors, and data frames. A *vector* is an ordered collection of values of the same type. A *matrix* is a two dimensional structure and an *array* can have even more dimensions. Their values are always of the same type. *Lists* allow storing of values that are of different types. *Factors* look like strings but represent categorical values internally stored as integers. *Data frames* store data in a tabular format; their individual rows represent data records and the columns hold their properties (the values in one column is always of the same type).

Unlike in some other programming languages, no data types need to be declared for variables. The data type of an object is given by what it contains. One object can hold a number and later, for example, a string.

## 2.6   Objects in R

All code in R manipulates *objects* [4]. *Object* is just something that is represented in a computer, like a vector, list, or data frame. It has a *class* that defines what functions can be used together with the object and how will they work. The other object classes besides the ones related to storing data include, for example, the `function` or `expression` classes.

```
> class(1)
[1] "numeric"
> class("hello")
[1] "character"
> class(class)
[1] "function"
> class(expression(1+2))
[1] "expression"
```

The objects usually have a name (known as a symbol) so we can later manipulate them. They thus behave like variables in other programming languages. The set of symbols is called the *environment* (which is also an object). The list of all objects in the current environment is returned by the objects() function. A new environment is created when a new session in R is started. A new environment is also created in the body of a function. The parent environment of the function is then the environment in which the function was created (not necessarily the same as where the function is called).

Variable names in R are case-sensitive so, for example, x is not the same as X. Variable names can contain alphanumeric characters, periods, and underscores, but they cannot start with a number or an underscore or with a period followed by a digit.

A list of existing objects in a specified environment (or objects defined by a user or local names in a function when called without parameters) is accessible through the ls() or objects() functions.

Variables can be easily removed using the remove() function (or using a shortcut rm()):

```
> x <- 1
> x
[1] 1
> remove(x)
> x
Error: object 'x' not found
```

Objects can have properties known as *attributes*. The list of attributes of an object is returned by the attributes() function. The attributes that are common to many object types include, for example, class – the class of an object, dim – dimensions of an object, names – the names of data elements (e.g., in a vector), levels – levels of a factor, among others.

To access or modify the value of an attribute, we usually use a function that has the same name as the attribute.

```
> x <- c(first=1, second=2, third=3)
> x
 first second  third
     1      2      3
> attributes(x)
$'names'
[1] "first"  "second" "third"

> names(x)
[1] "first"  "second" "third"
```

```
> names(x) <- c("1st", "2nd", "3rd")
> x
1st 2nd 3rd
  1   2   3
> names(x)
[1] "1st" "2nd" "3rd"
> attributes(x)
$'names'
[1] "1st" "2nd" "3rd"
```

There are several so called *generic functions*. Their names are shared across different classes and they implement different algorithms depending on the class (they enable polymorphism). There is, for example, a difference when printing a number and a character string, where the latter requires double quotes to appear around the value.

There are also many functions that enable the examination of the features of different objects, for example:

- class() – the class of an object (a vector of names of classes an object inherits from); if not known, generic classes, like matrix, array, function, numeric, or the result of typeof() are returned

- typeof() – the internal type of data stored in an object (e.g., integer or logical)

- length() – the length of an object, e.g., the number of vector elements

- attributes() – returns some metadata related to an object, like the dimensions of a matrix

- str() – compactly displays the internal structure of an object

- exists() – gives information about the existence of an object with a given name

- get() – returns an object with a name passed as the parameter

- print() – prints the values of its arguments

- summary() – gives some basic information about a data structure, for example, the minimum, maximum, mean, median, first quartile, and third quartile for numeric vectors or the levels and numbers of elements for each level in the case of factors.

To find out whether an object is of a given class, the is() function returning a logical value can be used:

```
> x <- 1
> is(x,"numeric")
[1] TRUE
> is(x,"character")
[1] FALSE
```

Many common classes have their own functions which can answer the same question, for instance is.numeric(), is.matrix(), is.factor() etc.

To change the class of an object (which is known as casting [58]), a class can be directly assigned to the object:

```
> x <- 1
> x
[1] 1
> class(x) <- "character"
> x
[1] "1"
```

This is, however, not recommended. Instead, a set of functions with the names starting with as. that are corresponding to the is.* functions is available and should be used.

```
> x <- 1
> x <- as.character(x)
> x
[1] "1"
```

## 2.6.1 Assignment

The assignment operator <- points to the object receiving the value of the expression stated on the right side. To create an object with value 10 we can write

```
> x <- 10
> x
[1] 10
```

In an assignment, the objects are usually immutable which means that their values are copied instead of assigning just a reference to them, as shown in the following example:

```
> a <- 1
> b <- a
> c <- a     # both 'b' and 'c' contain the value of 'a'
> a
[1] 1
> b
[1] 1
> c
[1] 1        # both 'b' and 'c' have the same value 1
> b <- 2
> b
[1] 2
> c
[1] 1        # the value of 'c' has not changed
> a <- 2
> c
[1] 1        # the value of 'c' still has not changed
```

There are a few other alternatives besides <−, such as =, −>, <<−, and −>>. It is, however, recommended that <− is used, because it always means assignment. Assignment of a value to a variable can be also achieved with using the assign() function (a variable name is passed as a character string):

```
> assign("x", 10)
```

## 2.6.2  Logical Values

Logical values can either be TRUE (when used as numbers, they are treated as 1) or FALSE (numerically 0). Shortcuts T and F can be used instead which is, however, not recommended since these names can be also used as variable names and thus easily overwritten.

```
> class(TRUE)
[1] "logical"
> 2*TRUE
[1] 2
> 2*FALSE
[1] 0
> T
[1] TRUE
> T == TRUE
```

```
[1] TRUE
> T <- 0
> T == TRUE
[1] FALSE
```

### 2.6.3 Numbers

Decimal values that are called numeric in R are the default computational data type. They include integers and decimals, positive, negative, and zero. The values are represented in double precision floating point. Integers which hold only whole numbers need to have an L appended at the end of the value.

```
> class(1)
[1] "numeric"
> class(1L)
[1] "integer"
> class(1.5)
[1] "numeric"
```

To create an integer variable in R, we can also call the as.integer function.

```
> class(1)
[1] "numeric"
> class(as.integer(1))
[1] "integer"
> class(as.integer(1.8))
[1] "integer"
> x <- as.integer(1.8)
> x
[1] 1
> class(x)
[1] "integer"
```

Optionally, numbers can be written using the semi-logarithmic notation.

```
> 1e2
[1] 100
> class(1e2)
```

```
[1] "numeric"
> 1e-2
[1] 0.01
```

You may also specify the value in the hexadecimal notation by prefixing the value with 0x.

```
> 0x1
[1] 1
> 0xff
[1] 255
> class(0xff)
[1] "numeric"
```

R can also work with complex numbers (written for example, as 1+1i) but they are not very common in machine learning so far.

There are a number of built-in functions for working with numbers. They represent typical operations, like the absolute value (abs()), square root ((sqrt())), logarithm (()), exponentiation (exp()), different forms of rounding (floor()), ceiling(), trunc(), round(), or signif()), trigonometric and inverse trigonometric functions (sin()), cos(), tan(), asin(), acos(), or atan()). Most of these functions work with one numeric vector (and are applied to its components). Some other functions require more arguments, for instance log(), that needs a base for the logarithm. Trigonometric functions in R measure angles in radians. The whole circle is $2\pi$ radians, a right angle $\pi/2$ radians. The value of the Ludolphian number $\pi$ is known as *pi* in R.

## 2.6.4 Character Strings

A data type enabling working with text data is character. Strings constants are enclosed between a pair of single or double quotes. Double quotes are preferred as strings are also quoted by double quotes on the output (to not quote them on the output, use the noquote() function). When a single or double quote is to appear in a string, the string might be enclosed between a pair of the other quote type:

```
> "This book is published by 'CRC Press'"
[1] "This book is published by 'CRC Press'"
> 'This book is published by "CRC Press"'
[1] "This book is published by \"CRC Press\""
```

The quotes can also be escaped by a backslash so they are not considered terminating quotes:

```
> "This book is published by \"CRC Press\""
[1] "This book is published by \"CRC Press\"'\n"
```

There are other symbols which can be inserted to strings using escaping characters by a backslash, examples being \n (newline), \r (carriage return), \t (tabulator), \b (backspace), \\ (backslash), \nnn (character with the given octal code consisting of 1 to 3 digits), \xnn (a character with the given hexadecimal code consisting of 1 or 2 digits), \unnnn (a Unicode character with the given code consisting of 1 to 4 hexadecimal digits), etc.

Since string data can be a part of one of the structures where R stores data, it is necessary to understand how they behave [237]:

■ vector, matrix – if one or more elements are strings, all data in a vector or matrix will be treated as strings

■ data frames – strings are converted to factors by default; to turn this behaviour off, the stringsAsFactors parameter should be set to FALSE in the data.frame() function while creating a data frame object

■ list – all elements maintain their types

There are many functions which can be used to handle string data. Here are a few typical examples:

■ nchar() – returns number of characters of a string[3]

■ tolower(), toupper() – returns a string converted to lower/upper case

■ substr(), substring() – returns or replaces substrings in character vectors

■ strsplit() – splits a string to substring according to the match of one or more regular expressions

## 2.6.5 Special Values

Missing values are represented by a constant NA (NA stands for *Not Available*). The is.na() or complete.cases() functions can be used to detect NA values in data structures.

---

[3]The function length(), that is, in many programming languages, used to give a string length, returns the number of elements in a data structure in R.

```
> x <- c(1, 2, 3, NA, 5)
> x
[1]  1  2  3 NA  5
> is.na(x)
[1] FALSE FALSE FALSE  TRUE FALSE
> complete.cases(x)
[1]  TRUE  TRUE  TRUE FALSE  TRUE
```

Calculations can sometimes lead to results that are either too big or too small to be stored in a memory. Very big (i.e., positive numbers) are represented by the Inf constant and very small (i.e., negative numbers) are represented by the −Inf constant.

```
> 2^1000
[1] 1.071509e+301
> 2^10000
[1] Inf
> -2^10000
[1] -Inf
```

To detect finite and infinite numbers, the is.finite() and is.infinite() functions can be used:

```
> x <- c(1, 1/0, -2^1000)
> x
[1]    1.000000e+00          Inf -1.071509e+301
> is.finite(x)
[1]  TRUE FALSE  TRUE
```

Other calculations can lead to results that cannot be represented as numbers. They are known as NaN (NaN stands for *Not a Number*). Basically all mathematical functions in R are supposed to work properly with NaN as input or output. To test whether a value is NaN, the is.nan() function is used.

```
> Inf/Inf
[1] NaN
> 0/0
[1] NaN
> Inf-Inf
[1] NaN
```

NULL is an object that represents an undefined object or a list with zero length (all NULLs point to the same object). It is not the same as missing value. It is often used as a parameter of a function when no value for that parameter should be passed or returned by a function when the value is undefined.

The function as.null() ignores its argument and returns NULL; is.null() tests whether the argument has an undefined value.

Dates are internally stored as numbers of days or seconds (depending on the class) since 1$^{st}$ January 1970. There are two date classes, Data, storing only information about days, and POSIXct, which also contains information about time. To create a date value, functions asData() or as.POSIXct() are used to convert strings to dates.

```
> date1 <- as.Date('2018-01-01')
> date1
[1] "2018-01-01"
> class(date1)
[1] "Date"
> typeof(date2)
[1] "double"
> date2 <- as.POSIXct('2018-01-01 10:20:30')
> date2
[1] "2018-01-01 10:20:30 CET"
> class(date2)
[1] "POSIXct" "POSIXt"
> typeof(date2)
[1] "double"
```

## 2.7   Functions

Functions are objects that accept a set of arguments, somehow process them, and return a value. They may, therefore, be used as parts of expressions.

Some functions are associated to some classes. They are known as methods. Unlike in other languages (like Java), not all functions are closely tied to a particular class in R [4].

Some of the methods have the same name in different classes. They are called *generic functions*. They enable the use of the same code for working with objects of different classes. The most visible generic function is print() which is automatically called after commands that are evaluated by R return any object. The objects are somehow printed depending on their type (e.g., a vector is printed as a list of whitespace separated values on a row, a matrix is printed as M rows and N columns).

There exist a large number of built-in functions from many categories which are readily available for use. They include functions enabling working with objects of different types of data (functions for numbers, character string, factors etc.), mathematical, statistical or optimization functions, functions for generating graphics, etc.

Other functions become available after a package containing their definitions is loaded.

Sometimes, a programmer might need a function that is not available and needs to create a new one. In that case, the algorithm the function implements needs to be created and the interface of the function must be specified. The interface includes the parameters and the value to be returned by the function. The body of the function is a code that usually processes input parameters to produce a return value. A function often has a name, which means that the function is stored in an object with that name.

To define a function, the `function` statement is called. Here, the parameters (arguments) of a function are defined, together with the function body.

The section with argument definitions contains the names of arguments so that their names can be used in the function body. The body of a function contains some statements to be executed. If more than one statement is contained in the body, they need to be enclosed within curly brackets. The return value, which can be an object of any type, is the value of the last evaluated expression. To explicitly return a value, the `return()` function is used.

```
> XtoN <- function(x, n) {
+ return(x^n)
+ }
> XtoN(2, 3)
[1] 8
```

The parameters can also have default values. The default values are used when real parameters are not supplied. Such parameters are also treated as optional.

```
> XtoN <- function(x, n=1) {
+ return(x^n)
+ }
> XtoN(2)
[1] 2
```

Parameters in R are often of a variable length. The list of such parameters is known as an object with a special name . . . (available in a function body).

```
> sum_all <- function(...) {
+     s <- 0;
+     for (x in list(...)) {
+         s <- s+x
+     }
+     return(s)
+ }
> sum_all(1, 2, 3, 4, 5)
[1] 15
```

Modifying a parameter inside a function does not change the value of the real parameter. The initial value of the parameter will be the one of the actual parameter, but its subsequent modifications will not affect the actual argument. In addition, a variable appearing within a function that has the same name as a variable that is global to it will be treated as local. This means that changing its value in a function will not affect the value of the global variable. Only its initial value will be that of the global [183].

To modify the value of an actual parameter or of a global variable, the superassignment operator $<<-$ is used.

```
> x <- 1
> add_one <- function(x) x <- x+1
> add_one(x)
> x
[1] 1
> add_one <- function(x) x <<- x+1
> add_one(x)
> x
[1] 2
```

Let's have a function XtoN() for calculating the value of base to the power of the exponent:

```
XtoN <- function(base, exponent) return(base^exponent)
```

When submitting actual parameters during when calling a function, the following possible methods, that are tried in the following order, can be used [213, 4]:

1. The parameters are submitted as pairs *full_name=value*. Then, the order of parameters does not matter.

```
> XtoN(base=2, exponent=3)
[1] 8
> XtoN(exponent=3, base=2)
[1] 8
```

2. The parameters are submitted as pairs *name=value*, where *name* can be only a part of a parameter name. The order of parameters is also not important.

```
> XtoN(exp=3, b=2)
[1] 8
```

3. The values are passed in the same order as in the definition of the function. The first parameter will be stored to x and the second to n.

```
> XtoN(2, 3)
[1] 8
```

Information about the parameters of a function can be obtained from the documentation or by calling the args() function:

```
> args(XtoN)
function (base, exponent)
NULL
> args(sum_all)
function (...)
NULL
```

Functions can be also passed as arguments to other functions. In this case, they are often anonymous functions, which means that they do not have a name.

```
> myfunction <- function(x, f) { return(f(x)) }
> myfunction(c(5, 2, 3, 4, 8), min)
[1] 2
> myfunction(c(5, 2, 3, 4, 8), max)
[1] 8
> myfunction(c(5, 2, 3, 4, 8),
+            function (x) {return(length(x))}
+            )
[1] 5
```

## 2.8 Operators

An operator is a function working with one or two parameters which do not require writing parentheses. Some of the basic operators, which have not been discussed in much detail so far, are listed in Table 2.1. A user can also write his own binary operators. They are named %operator_name% and are implemented through functions.

**Table 2.1:** Some commonly used operators in R.

| | |
|---|---|
| *Arithmetic Operators* | |
| + | addition |
| − | subtraction (binary) or arithmetic negation (unary) |
| * | multiplication |
| / | division |
| %% | the modulo operation (the remainder after division of one number by another) |
| %/% | integer division |
| ^ | exponentiation (raising to a power) |
| *Relational Operators* | |
| <, <= | less than, less or equal |
| >, >= | greater than, greater or equal |
| == | equal to |
| != | not equal to |
| *Logical Operators* | |
| ! | negation |
| \| | logical disjunction (OR), processes all vector elements |
| \|\| | logical disjunction (OR), works only with the first vector element |
| & | logical conjunction (AND), processes all vector elements |
| && | logical conjunction (AND), works only with the first vector element |
| *Colon Operator* | |
| : | sequence of numeric values |

In order to write expressions so that they return a correct value, the precedence (priority) of operators is important. The following list contains a complete set of operators in R, placed according to their priority in descending order [260]:

■ function calls and groupings (( {)

■ access variables in a namespace (:: :::)

■ component / slot extraction ($ @)

■ indexing ([ [[)

- exponentiation (right to left) (^)

- unary plus and minus (+ -)

- sequence operator (:)

- special operators (%any%, including %% and %/%)

- multiplication, division (* /)

- addition, subtraction (+ -)

- comparison (< > <= >= == !=)

- negation (!)

- logical conjunction (& &&)

- logical disjunction (| ||)

- separating the left and right-hand sides in a model formula (~)

- rightwards assignment (-> ->>)

- assignment (right to left) (<- <<-)

- assignment (right to left) (=)

- help (?)

## 2.9 Vectors

The simplest data structure in R is a *vector*. It is an ordered collection of values of the same type (they are known as atomic structures). We can therefore distinguish, for example, numeric (integer and double-precision), logical (containing the values TRUE, FALSE, and NA), and character vectors. Even a single number in an expression (like 1+2) is treated as a numeric vector with length 1.

### 2.9.1 Creating Vectors

To create a vector, the function c() is typically used. The function takes its parameters and creates a vector from them. If the parameter values are of different types they are coerced into one type (c might thus stand for coerce).

```
> x <- c(1, 2, 3, 4, 5)
> x
[1] 1 2 3 4 5
> x <- c(1, 2, 3.5)
> x
[1] 1.0 2.0 3.5
> x <- c(1, 2, 3, "a", "b")
> x
[1] "1" "2" "3" "a" "b"
> x <- c(1, 2, 3, TRUE)
> x
[1] 1 2 3 1
```

If you forget to write the name of function c(), which is a common problem in R beginners, you will get an error.

```
x <- (1, 2, 3)
Error: unexpected ',' in "x <- (1,"
> cummin(1, 2, 3)
Error in cummin(1, 2, 3) :
  3 arguments passed to 'cummin' which requires 1
```

A vector can be also created using the vector() function. The first parameter (mode) specifies the type of the vector (numeric, logical, character, list, or expression). The second parameter (length) defines the number of elements to be created. All elements will have a value that means nothing (0 for numbers, " " for characters, FALSE for logical etc.).

The colon operator (:) can be used to create numeric sequences too. It requires two values. If the first is less than the second, an ascending sequence of numbers starting with the first value and ending with the second value with step 1 will be created. If the first value is greater than the second, a descending sequence will be produced.

```
> x <- 1:5
> x
[1] 1 2 3 4 5
> x <- 5:1
> x
[1] 5 4 3 2 1
```

A more flexible way to create sequences is using the seq() function. It accepts a few parameters (not all of them are mandatory): from – the starting value of a sequence, to – the end value of the sequence, by – an increment of the sequence, length.out – a desired length of the sequence, along.with – an object from which the length of the sequence is taken.

```
> x <- seq(1, 5)
> x
[1] 1 2 3 4 5
> x <- seq(-1, 1, by=0.3)
> x
[1] -1.0 -0.7 -0.4 -0.1  0.2  0.5  0.8
> x <- seq(-1, by=0.3, length=3)
> x
> y <- 1:3
> x <- seq(1, 5, along.with=j)
> x
[1] 1 3 5
```

Another useful function is rep(). It replicates an object according to the passed arguments which include: times – how many times the values should be repeated, length.out – what should be the length of the sequence, and each – how many times each element will be repeated.

```
> x <- c(1, 2, 3)
> rep(x)
[1] 1 2 3
> rep(x, times=2)
[1] 1 2 3 1 2 3
> rep(x, times=2, length.out=10)
[1] 1 2 3 1 2 3 1 2 3 1
> rep(x, length.out=10)
[1] 1 2 3 1 2 3 1 2 3 1
> rep(x, length.out=10, each=4)
[1] 1 1 1 1 2 2 2 2 3 3
```

## 2.9.2  Naming Vector Elements

In some situations, it is more convenient to access vector values using some names. They can usually be remembered more easily than numbers (positions).

Fortunately, R provides a mechanism of naming vector elements. The names can be defined at the moment of creating a vector. The name (if it looks like a regular name, it does not have to be quoted), together with a equal sign, precedes a value. On the output, the names will be displayed above the values of the vector.

```
> x <- c(first=1, second=2, third=3)
> x
 first second  third
     1      2      3
```

Not all values need to have names.

```
> x <- c(first=1, 2, third=3)
> x
 first         third
     1      2      3
```

The names of a list are returned by the names() function. When a vector has no element names, NULL is returned.

```
> x <- c(first=1, second=2, third=3)
> names(x)
[1] "first"  "second" "third"
```

The names can be at any time modified using the names attribute of a vector (through the names() function).

```
> day_names <- c('Mon', 'Tue', 'Wed', 'Thu', 'Fri',
+                'Sat','Sun')
> x <- 1:7
> names(x) <- day_names
> x
Mon Tue Wed Thu Fri Sat Sun
  1   2   3   4   5   6   7
```

When enough values are not provided for the names, their value will be NA. An empty name is simply an empty string.

```
> x <- 1:5
> names(x) <- c("a", "b", "", "d")
> x
   a   b       d <NA>
   1   2   3   4   5
```

### 2.9.3 Operations with Vectors

When vectors are used in arithmetic expressions, the operations are performed on the level of individual vector elements. The result has the same length as the longest vector. The values of the shorter vector are repeated (recycled) so there are enough values as needed for all the operations.

```
> x <- c(1, 2, 3)
> 2*x
[1] 2 4 6
> x <- c(1, 1, 1, 1, 1, 1)
> y <- c(1, 2)
> x+y
[1] 2 3 2 3 2 3
```

There is a similar situation in the case of logical operations. The operation is performed with each element of a vector and the result will be a logical vector with the length of the original vector.

```
> x <- 1:5
> y <- x>=3
> y
[1] FALSE FALSE  TRUE  TRUE  TRUE
> !y
[1]  TRUE  TRUE FALSE FALSE FALSE
```

There are many useful functions operating on vectors, like min() (finds the minimum), max() (finds the maximum), range() (finds both extremes), sum() (calculates the sum of values), prod() (calculates the product of all values), mean() (returns the mean), var() (calculates the variance), sort() (sorts the values), order() (returns the order of sorted values in the input), or length() (counts the elements of a vector).

```
> min(c(5, 3, 4, 9, 1, 2))
[1] 1
> max(c(5, 3, 4, 9, 1, 2))
[1] 9
> range(c(5, 3, 4, 9, 1, 2))
[1] 1 9
> sum(c(5, 3, 4, 9, 1, 2))
[1] 24
> prod(c(5, 3, 4, 9, 1, 2))
[1] 1080
> 5*3*4*9*1*2
[1] 1080
> mean(c(5, 3, 4, 9, 1, 2))
[1] 4
> var(c(5, 3, 4, 9, 1, 2))
[1] 8
> length(c(5, 3, 4, 9, 1, 2))
[1] 6
> sort(c(5, 3, 4, 9, 1, 2))
[1] 1 2 3 4 5 9
> order(c(5, 3, 4, 9, 1, 2))
[1] 5 6 2 3 1 4
```

The vector length can be changed using the length attribute.

```
> x <- 1:10
> length(x) <- 5
> x
[1] 1 2 3 4 5
> x <- 1:10
> x
[1]  1  2  3  4  5  6  7  8  9 10
> length(x) <- 5
> x
[1] 1 2 3 4 5
> length(x) <- 10
> x
[1]  1  2  3  4  5 NA NA NA NA NA
```

The paste() function takes an arbitrary number of character vectors and concatenates them, one by one, into one vector. All non-strings will be coerced into character strings. The `sep` parameter is used to connect individual concate-nated strings (a space is used by default). If the `collapse` parameter is provided, all strings from the resulting vector are concatenated to a single string where the value of the `collapse` parameter is used to join the elements.

```
> paste(c("a", "b", "c"), 1:3, c("x", "y", "z"))
[1] "a 1 x" "b 2 y" "c 3 z"
> paste(c("a", "b"), rep(1:5, each=2), sep="-")
[1] "a-1" "b-1" "a-2" "b-2" "a-3" "b-3" "a-4" "b-4"
[9] "a-5" "b-5"
> paste(c("a", "b", "c"), 1:3, sep="", collapse="+")
[1] "a1+b2+c3"
```

## 2.9.4 Accessing Vector Elements

Some of the elements of a vector may be accessed using an integer vector known as an *index vector* (non-integer values are converted to integers by truncating the values). The index vector, which is written into a pair of square brackets, contains the positions of elements that are to be processed. The elements are counted from 1.

The following example shows how to retrieve a vector member or several members.

```
> x <- 10:15
> x
[1] 10 11 12 13 14 15
> x[2]
[1] 11
x[2:4]
[1] 11 12 13
> x[2.4]
[1] 11
> x[2.7]
[1] 11
```

If an index is out of the range of the vector, a missing value (NA) will be returned.

```
> x[c(1, 20)]
[1] 10 NA
```

Using a logical vector as the index means that only the elements which correspond to the positions with TRUE values of the index will be included in the output. Both vectors should normally have the same length. If the index vector is shorter, its elements will be recycled.

```
> x[c(TRUE, FALSE, FALSE, TRUE)]
[1] 10 13 14
```

If the index values are negative they are understood as positions of vector elements to be excluded.

```
> x
[1] 10 11 12 13 14 15
> x[-1]
[1] 11 12 13 14 15
> x[-seq(from=2, by=2, along.with=x)]
[1] 10 12 14
```

When the elements of a vector have names and the index vector contains strings, elements corresponding to the names are returned. When no names exist or when a non-existing name is supplied the corresponding elements will have the NA value.

```
> x <- c(first=1, second=2, third=3)
> x["first"]
first
    1
> x["firs"]
<NA>
   NA
```

## 2.10   Matrices and Arrays

*Matrices* are two-dimensional objects that, similarly to vectors, contain values of the same type (numbers, characters, logical values). To create a matrix, the matrix() function might be used. A matrix is created from some object containing data (passed as the first parameter or a parameter named data) and is organized into two dimensions according to the parameters of the function. The parameters include: `nrow` — the number of rows, `ncol` – the number of columns, `byrow` – when FALSE, which is a default option, the matrix is filled by columns, otherwise by rows, and `dimnames` – a list with the names assigned to the rows and columns.

```
> matrix(1:9, nrow=3)   # same as matrix(data=1:9, nrow=3)
     [,1] [,2] [,3]
[1,]   1    4    7
[2,]   2    5    8
[3,]   3    6    9
> matrix(1:9, nrow=3, byrow=TRUE)
     [,1] [,2] [,3]
[1,]   1    2    3
[2,]   4    5    6
[3,]   7    8    9
> rnames <- c("x1", "x2", "x3")
> cnames <- c("y1", "y2", "y3")
> matrix(1:9, nrow=3, byrow=TRUE,
+          dimnames=list(rnames, cnames))
   y1 y2 y3
x1  1  2  3
x2  4  5  6
x3  7  8  9
```

Information about matrix dimensions can be obtained using the `nrow`, `ncol`, and `dim` functions.

```
> x <- matrix(1:10, nrow=5)
> nrow(x)
[1] 5
> ncol(x)
[1] 2
> dim(x)
[1] 5 2
> length(x)
[1] 10
```

To access individual elements of a matrix, two values of an index are needed. They are again written between a pair of square brackets. When accessing a position out of the range of the two indexes, an error is caused.

```
> x <- matrix(1:10, nrow=5, byrow=TRUE)
> x[2, 1]
[1] 3
> x[2, 4]
Error in x[2, 4] : subscript out of bounds
```

The entire row of a matrix can be accessed by leaving the second index empty. When we want to access a column, the row index is not entered.

```
> x <- matrix(1:10, nrow=5, byrow=TRUE)
> x[2, ]
[1] 3 4
> x[, 1]
[1] 1 3 5 7 9
```

When the indices for a row, a column, or both, contain a vector, more than one rows or columns are accessed at once. The same principles apply as when accessing vector elements (you might use, for example, a logical vector).

```
> x[c(1, 3, 5), ]
     [,1] [,2]
[1,]   1    2
[2,]   5    6
[3,]   9   10
> x[c(1, 3, 5), c(FALSE, TRUE)]
[1]  2  6 10
```

Arithmetic operations with matrices work element-wise.

```
> m1 <- matrix(rep(2, 4), nrow=2)
> m2 <- matrix(rep(3, 4), nrow=2)
> m1+m2
     [,1] [,2]
[1,]   5    5
[2,]   5    5
> m2-m1
     [,1] [,2]
[1,]   1    1
[2,]   1    1
```

```
> m2*m1
     [,1] [,2]
[1,]    6    6
[2,]    6    6
> m1/m2
            [,1]       [,2]
[1,] 0.6666667 0.6666667
[2,] 0.6666667 0.6666667
> sin(m1)
            [,1]       [,2]
[1,] 0.9092974 0.9092974
[2,] 0.9092974 0.9092974
```

An *array* is an object which can store data in more than two dimensions. In fact, arrays are very similar to matrices (arrays contain matrices – for example, a three dimensional array is like a "list" containing two dimensional matrices). When creating arrays or when accessing the elements of parts of arrays, we need more indices. An array is created using the array() function.

```
> x <- array(1:18, dim=c(3, 2, 3))
> x
, , 1
     [,1] [,2]
[1,]    1    4
[2,]    2    5
[3,]    3    6
, , 2
     [,1] [,2]
[1,]    7   10
[2,]    8   11
[3,]    9   12
, , 3
     [,1] [,2]
[1,]   13   16
[2,]   14   17
[3,]   15   18
> x[, , 1]
     [,1] [,2]
[1,]    1    4
[2,]    2    5
[3,]    3    6
> class(x[, , 1])
```

```
[1] "matrix"
> x[, 1, 1]
[1] 1 2 3
> x[1, 1, ]
[1]  1  7 13
```

## 2.11   Lists

A *list* is an object that can hold data elements of different types. It is, therefore, a heterogeneous structure that is very common in many situations. It can contain vectors, matrices, arrays, lists, or even functions. The list() function is used to create a list. A list is created from the objects that are passed as parameters similar to the case of vectors. Names can be assigned to individual elements during the creation of a list or later, using the names() function.

```
> person <- list("John", "Smith", c("Peter", "Paul"), TRUE)
> person
[[1]]
[1] "John"

[[2]]
[1] "Smith"

[[3]]
[1] "Peter" "Paul"

[[4]]
[1] TRUE

> person <- list(name="John", surname="Smith",
+                children=c("Peter", "Paul"),
+                insurance=TRUE)
> person
$`name`
[1] "John"

$surname
[1] "Smith"

$children
[1] "Peter" "Paul"
```

```
$insurance
[1] TRUE
```

The length of a list is the number of top-level elements.

```
> length(person)
[1] 4
```

Since lists can contain objects of many different types, they have no dimensions. It does not make sense either, to carry out arithmetic operations upon lists (e.g., sum two lists). The calculations can, of course, be made on the level of individual list elements.

To access list elements, the same way of indexing, as in the case of vectors, is used. We can use numeric (both positive and negative) or logical indexes or element names written within a pair of square brackets. The indexing operation using single square brackets always returns a list.

```
> person[1]
$'name'
[1] "John"

> person["children"]
$'children'
[1] "Peter" "Paul"

> class(person["insurance"])
[1] "list"
```

When we need to access the content of a list element, we write the number or name of the element between a pair of double square brackets.

```
> person[[2]]
[1] "Smith"
> person[["children"]]
[1] "Peter" "Paul"
> class(person[["insurance"]])
[1] "logical"
```

The same can be achieved using the notation of the dollar sign after a list name, followed by an element name. The advantage of this notation is that in many R environments, the item names are automatically filled and partial matches are also accepted.

```
> person$surname
[1] "Smith"
> person$child
[1] "Peter" "Paul"
```

## 2.12   Factors

*Factors* are used to represent data that can have a finite number of values (we can say that they are categorical or enumerated). A typical example is of class labels in a classification problem. The labels look like strings but, in fact, represent the categories of data. The different values of a factor are known as *levels*. To create a factor from a vector, the `factor()` function is used. The `levels()` function returns the levels of a factor.

```
> labels <- c("negative", "neutral", "positive",
+             "neutral", "positive")
> labels
[1] "negative" "neutral"  "positive" "neutral"  "positive"
> factor(labels)
[1] negative neutral  positive neutral  positive
Levels: negative neutral positive
> nlevels(factor(labels))
[1] 3
```

When creating a factor, the levels allowed can be supplied using the `levels` parameter; excluded levels are listed in the `exclude` parameter (other values will be transformed to `NA`). The level names can be defined with the `labels` parameter which also enables mapping multiple values to one level.

Internally, factors are stored as vectors of integer values where the corresponding labels are mapped to them. This is a great improvement in efficiency, since storing integers is much more efficient than storing strings. Normally, the order of factor values is not important. In some situations, however, such an ordering makes sense (for example, positive is more than negative, February is less than April).

The order of levels is given by the order of values in the `levels` parameter of the `factor()` function (when not provided, the levels are not ordered and are displayed in an alphabetical order). Ordering is visualized in the output of a factor. The ordering can be later changed by assigning new levels to the `levels()` function or calling the `factor()` function with an existing factor as a parameter together with new levels.

```
> f <- factor(labels,
+             levels=c("positive", "neutral", "negative"),
+             ordered=TRUE)
> f
[1] negative neutral  positive neutral  positive
Levels: positive < neutral < negative
> f <- factor((f), levels=c("negative", "neutral", "positive"))
> # the same as above
> levels(f) <- c("negative", "neutral", "positive")
> f
[1] positive neutral  negative neutral  negative
Levels: negative < neutral < positive
```

Ordering factor levels is important when we want, for example, to compare factor values.

```
> f
[1] positive neutral  negative neutral  negative
Levels: negative < neutral < positive
> ff <- f
> ff <- factor(f, ordered=FALSE)
> ff
[1] positive neutral  negative neutral  negative
Levels: negative neutral positive
> f[1]<f[2]
[1] FALSE
> ff[1]<ff[2]
[1] NA
Warning message:
In Ops.factor(ff[1], ff[2]) : < not meaningful for factors
```

To get rid of unused factor levels (when there are more levels defined than in the data), the `droplevels()` function can be used.

To access individual factor elements, an index can be used in the same manner as in case of vectors (positive and negative integers, logical values, and names).

```
> f[1]
[1] positive
Levels: negative < neutral < positive
> f[c(-1, -3)]
[1] neutral  neutral  negative
Levels: negative < neutral < positive
> f[c(TRUE, FALSE, TRUE)]
[1] positive negative neutral
Levels: negative < neutral < positive
```

The elements of a factor can also be modified using, for example an assignment. When we try to assign a value that is not within the allowed levels, an NA value will be inserted instead. To prevent this problem, the new level needs to be added.

```
> f[2] <- "positive"
> f
[1] positive positive negative neutral  negative
Levels: negative < neutral < positive
> f[2] <- "strongly positive"
Warning message:
In '[<-.factor'('*tmp*', 2, value = "strongly positive") :
  invalid factor level, NA generated
> f
[1] positive <NA>     negative neutral  negative
> levels(f) <- c(levels(f), "strongly positive")
> f[2] <- "strongly positive"
> f
[1] positive strongly positive negative neutral negative
Levels: negative < neutral < positive < strongly positive
```

## 2.13   Data Frames

In many situations, it is necessary to store records of data where individual elements of these records have different data types. In a scientific context, various experiments typically lead to a set of individual observations, each of which is described by the same set of measurements. This situation is also typical in relational databases and spreadsheets that both store data in a tabular format. Every rows represents one record and the columns contain their properties. The values in every column are of the same type and all rows have the same number of data elements.

*Data frames* are used to store such data in R. To create a data frame, the values to be contained in the columns are provided as parameters. Each column must have the same number of elements or else, as many as can be recycled by an exact number of times (for example, if one column has 3 values, the other 4, and the last 12, everything is fine because the first column is used four times, the second three times, and the last one once).

The columns usually have names. When they are not provided, R creates some column names on its own. In order to enable duplicate column names, the check.names parameter needs to be set to FALSE. Alternatively, names for the rows can be provided in the parameter row.names or taken from any of the parameters that has names. When names are not provided, the rows are numbered.

```
> data.frame(c("a", "b"), c(1, 2))
  c..a....b.. c.1..2.
1           a       1
2           b       2
> data.frame(column1 = c("a", "b"), column2 = c(1, 2))
  column1 column2
1       a       1
2       b       2
> names <- c('John', 'Paul', 'Peter')
> ages <- c(20, 50, 25)
> children <- c(TRUE, FALSE, TRUE)
> df <- data.frame(names, ages, children)
> df
  names ages children
1  John   20     TRUE
2  Paul   50    FALSE
3 Peter   25     TRUE
> df <- data.frame(names, ages, children,
+                  row.names=c("A", "B", "C"))
> df
  names ages children
A  John   20     TRUE
B  Paul   50    FALSE
C Peter   25     TRUE
```

When a character vector is used as a part of a data frame the values are automatically converted to factors. To prevent this, the `stringsAsFactors` parameter needs to be set to FALSE.

```
> d <- data.frame(a=c("a", "b"))
> class(d$a)
[1] "factor"
> d <- data.frame(a=c("a", "b"), stringsAsFactors=FALSE)
> class(d$a)
[1] "character"
```

Data frames are implemented as lists where the columns are list elements. Thus, the `length()` function for a data frame returns the number of columns.

To extract individual columns from a data frame, we can use the same principles as when accessing lists. A notation using the number or name of a column within square brackets is possible.

```
> df[1]
  names
A  John
B  Paul
C Peter
> df["names"]
  names
A  John
B  Paul
C Peter
> df[c("ages","names")]
  ages names
A   20  John
B   50  Paul
C   25 Peter
```

The returned values are again lists. To access the content, a double square bracket or dollar notation can be used.

```
> df[["ages"]]
[1] 20 50 25
> df$ages
[1] 20 50 25
```

```
> class(df["ages"])
[1] "data.frame"
> class(df[["ages"]])
[1] "numeric"
> class(df$ages)
[1] "numeric"
```

More frequently, a pair of vectors containing indices (positive or negative integers, names, or logical values) is used to specify which rows and columns to extract. The first list refers to the extracted rows, the second vector specifies the columns.

```
> df[1:2, c("names", "children")]
  names children
A  John     TRUE
B  Paul    FALSE
> df[, c("names", "children")]
  names children
A  John     TRUE
B  Paul    FALSE
C Peter     TRUE
> df[1:2, ]
  names ages children
A  John   20     TRUE
B  Paul   50    FALSE
```

The returned value is, again, a data frame because severals rows consisting of columns of different data types are generally obtained. Only in the case that one column is returned, is the value is vector.

```
> class(df[1:2, ])
[1] "data.frame"
> class(df[, 2])
[1] "numeric"
```

To add a new column to a data frame or to replace the values of an existing column, a vector of values can be easily assigned to an appropriate index (using a number or name). An alternative is the cbind() function which combines two data frames so the resulting data frame contains the columns from both data frames (duplicate columns are not checked). The following commands have the same result.

```
> df[4] <- c("New York", "Boston", "Chicago")
> df[[4]] <- c("New York", "Boston", "Chicago")
> df["city"] <- c("New York", "Boston", "Chicago")
> df[["city"]] <- c("New York", "Boston", "Chicago")
> df$city <- c("New York", "Boston", "Chicago")
> df <- cbind(df,
+         data.frame(city=c("New York", "Boston", "Chicago")))
```

To delete a column, the NULL value is assigned to it.

```
> df["city"] <- NULL
```

When we need to add a row (or multiple rows) to a data frame, we actually have to merge two data frames. Both data frames must have the same column names (the order does not have to be the same).

```
> df <- rbind(df,
+             data.frame(ages=60,
+                        names="George",
+                        children=TRUE,
+                        city="Los Angeles",
+                        row.names="D"))
> df
    names ages children        city
A    John   20     TRUE    New York
B    Paul   50    FALSE      Boston
C   Peter   25     TRUE     Chicago
D  George   60     TRUE Los Angeles
```

## 2.14   Functions Useful in Machine Learning

Besides the functions that have been mentioned in relation with objects, their classes, data types, basic operations with numbers, strings, vectors, matrices, factors, etc., R provides many built in functions which can be useful in a text mining process.

Let there be a matrix representing the frequencies of words in documents from three classes. To make the example simple, only nine different words are considered to be important. There are ten documents from classes labeled with numbers 1, 2, and 3.

```
> # creating a data frame  representing the distribution
> # of words in documents from three classes
> d <- data.frame(matrix(
+  c(3,0,0,0,1,0,0,0,1,1,
+    0,1,1,1,0,0,1,0,0,1,
+    0,0,1,0,1,0,0,1,0,1,
+    0,0,1,0,1,0,0,1,0,1,
+    1,0,0,2,1,1,0,0,0,2,
+    0,0,0,1,0,1,0,0,1,2,
+    0,0,0,0,1,2,1,0,0,2,
+    1,0,0,0,0,0,3,0,2,3,
+    0,0,0,1,0,0,1,2,1,3,
+    0,0,2,0,1,0,2,0,1,3),
+  ncol=10,byrow=TRUE))

> colnames(d) <- c("football", "hockey", "run",
+                   "president", "country", "Germany",
+                   "software", "device", "communication",
+                   "class")
```

To obtain some statistical information about the frequency of word Germany in documents (maximum, minimum, mean, etc.), we can call the summary() function.

```
> summary(d["Germany"])
    Germany
 Min.    :0.00
 1st Qu.:0.00
 Median :0.00
 Mean    :0.40
 3rd Qu.:0.75
 Max.    :2.00
```

To see how frequent are words in documents on average, means of individual columns can be calculated using the colMeans() function.

```
> # obtaining average frequencies of all words
> colMeans(d[, -10])
      football        hockey           run      president
           0.5           0.1           0.5           0.5
       country        Germany      software         device
           0.5           0.4           0.8           0.3
 communication
           0.6
```

Information about the typical document length and the variability of this property can be calculated using the functions mean() (arithmetic mean), var() (variance), and sd() (standard deviation).

```
> # obtaining the average length of a document
> # (not considering class labels)
> mean(rowSums(d[, -10]))
[1] 4.4
> var(rowSums(d[, -10]))
[1] 1.377778
> sd(rowSums(d[, -10]))
[1] 1.173788
```

Word frequencies and the distribution of documents in classes can be easily visualized (for more details about graphics, see 2.17).

```
> # visualizing global frequencies of all words
barplot(sort(colSums(d[, -10]), decreasing=TRUE))

> # a pie chart relative numbers of documents in classes
> pie(prop.table(table(d["class"])),
+      labels=paste("class",
+                 names(prop.table(table(d["class"]))),
+                 ":",prop.table(table(d["class"]))))
```

To see how individual pieces of data are related to each other, the function, table() can be used to create a contingency table of the counts at each combination of factor levels. It requires one or more objects which can be interpreted as factors, or a list (or data frame) whose components can be so interpreted. In our example, we can observe, how often the words *football* and *president* or word *run* and individual class labels co-occur.

```
> # how often words co-occur displayed in a contingency table
> table(lapply(d[, c("football","president")],
+         function (x) x>0))
         president
football FALSE TRUE
    FALSE     4    3
    TRUE      2    1


> # how often a word and class label co-occur displayed
> # in a contingency table
> table(d[,"class"], d[,"run"] > 0,
+         dnn=c("Class label", "Contains 'run'"))
            Contains 'run'
Class label FALSE TRUE
          1     1    3
          2     3    0
          3     2    1
```

Sometimes, it might be useful to find whether there are duplicates in the data. Logical values denoting indices of duplicated items in vectors or data frames (data frame rows) are returned by the function duplicated(). The supplied data structure is scanned according to the value of the parameter fromLast (FALSE by default) from the first or last element. If a current value has been found before, its index is marked as TRUE. The logical values can be used to index a structure to get the duplicates. If the values are negated, only the unique values can be obtained.

```
> # getting indices of duplicated items
> duplicated(d)
 [1] FALSE FALSE FALSE  TRUE FALSE FALSE FALSE FALSE
 [9] FALSE FALSE
> # removing the duplicates
> d <- d[!duplicated(d), ]
```

The function cut() is useful when the values of a numeric vector need to be divided into intervals and coded according to the interval in which they fall. The leftmost interval has level one, the next leftmost level two and so on. The values can be assigned to the given number of intervals or interval boundaries which are supplied (breaks parameter). If labels are provided, they are used to code the intervals instead of numbers. In our example, class label 1 will be converted to sport, the remaining labels to other.

```
> cut(d$class,
+      breaks=c(0, 1, 3),
+      labels=c("sport", "other"))
[1] sport sport sport other other other other other other
Levels: sport other
> # replacing numeric class labels with strings
> d$class = cut(d$class,
+                  breaks=c(0, 1, 3),
+                  labels=c("sport", "other"))
```

For some tasks, working only with a part of a data set might be needed. An example is *bagging* (Bootstrap aggregating), a method for creating an ensemble of classifiers trained on samples of data [32]. Another situation is when there are so many data instances available so that they cannot fit into the memory to be processed.

Sampling can be achieved using function sample(). It requires a vector of elements from which some will be chosen, or an integer (a sample will then be created from values 1:x). The number of elements is given by the size parameter. The elements can appear once (sampling without replacement, parameter replace is set to FALSE) or more than once in the selection (sampling with replacement, parameter replace is set to TRUE).

```
> # creating a sample containing 50% instances from
> # the entire data set
> d[sample(nrow(d), size=nrow(d)*0.5, replace=FALSE),]
  football hockey run president country Germany
3        0      0   1        0       1       0
9        0      0   0        1       0       0
6        0      0   0        1       0       1
1        3      0   0        0       1       0
  software device communication class
3        0      1             0 sport
9        1      2             1 other
6        0      0             1 other
1        0      0             1 sport
```

The function sample() can be easily used to randomize rows in a data frame. Shuffling, for example, training data after a few epochs of training can improve the performance of a learning system [51].

```
> # shuffling rows of d
> d <- d[sample(nrow(d)),]
```

Often, a function (transformation) needs to be applied on elements of a data structure (raising numbers in a vectors to a power of two, calculating sums of rows in a matrix, etc.). Instead of using loops to iterate over the structures, function apply() and other similar functions are a good choice.

The function apply() takes an array when an object of another type is supplied; R tries to coerce it to a matrix using as.matrix() and applies a function to its elements. The elements are defined by the second parameter (MARGIN), which, in the case of matrices, is 1 for rows, 2 for columns, c(1, 2) for both rows and columns (i.e., for each matrix element), or a list of dimension names when the dimensions have names. The function that is supplied as the third parameter (FUN) can be one of existing functions or a user defined one. Function apply() returns a vector or array or list of values according to the type of data returned by the supplied function.

```
> # counting the numbers of words in every document
> # (applying function apply() to rows)
> apply(d[, -10], 1, sum)
10  5  1  6  8  3  9  7  2
 6  5  5  3  6  3  5  4  4
```

```
> # counting the frequencies of words in all documents
> # (applying function apply() to columns)
> apply(d[, -10], 2, sum)
    football        hockey           run
           5             1             4
   president       country       Germany
           5             5             4
    software        device communication
           8             3             6
```

```
> # a user defined function is used to threshold frequencies
> # of words in documents (if a frequency is greater than 1
> # it is set to 2)
> apply(d[, -10], c(1, 2), function (x) if (x > 1) 2 else x)
```

|    | football | hockey | run | president | country | Germany |
|----|----------|--------|-----|-----------|---------|---------|
| 10 | 0 | 0 | 2 | 0 | 1 | 0 |
| 5  | 1 | 0 | 0 | 2 | 1 | 1 |
| 1  | 2 | 0 | 0 | 0 | 1 | 0 |
| 6  | 0 | 0 | 0 | 1 | 0 | 1 |
| 8  | 1 | 0 | 0 | 0 | 0 | 0 |
| 3  | 0 | 0 | 1 | 0 | 1 | 0 |
| 9  | 0 | 0 | 0 | 1 | 0 | 0 |
| 7  | 0 | 0 | 0 | 0 | 1 | 2 |

```
2         0      1    1            1        0      0
     software device communication
10        2      0                 1
5         0      0                 0
1         0      0                 1
6         0      0                 1
8         2      0                 2
3         0      1                 0
9         1      2                 1
7         1      0                 0
2         1      0                 0
```

Function `lapply()` is able to apply a supplied function to all elements of the passed object. In comparison to `apply()`, it does not require the `MARGIN` parameter. It works with vectors or lists (other objects are coerced to lists) and returns a list of the same length as that of the object. The same functionality is provided by the `sapply()` function that returns a vector.

Function `which()` returns the position where a logical object has TRUE values. If the logical object is an array, indices can be returned instead of scalar positions (parameter `arr.ind` must be set to TRUE).

```
> # obtaining positions of documents from class 2
> which(d$class == "sport")
[1] 3 6 9
```

## 2.15 Flow Control Structures

R provides a variety of control flow structures that are typical in many other programming languages. The flow control statements are all reserved words. A part of these statements is sometimes a condition based on which the code is executed. A *condition* is a logical vector of length one. If a longer vector is provided, only the first element will be processed.

### 2.15.1 Conditional Statement

The *conditional statement* is used when we need to decide whether to execute a statement or not or in the case where one of two possible statements is to be executed. The statement has a common form:

if (*condition*) *expression*

or, alternatively:

if (*condition*) *expression 1* else *expression 2*

When the condition is evaluated as TRUE, *expression 1* is evaluated. If the condition is evaluated as FALSE, nothing happens. If the *else* statement exists, *expression 2* will be evaluated in this case.

```
> x <- 2
> y <- 1
> if (y != 0) x/y else "Cannot divide by 0"
[1] 2
```

If an expression to be evaluated consists of more than one statement, the statements must be surrounded by curly braces. It is important to note that the else keyword must be on the same line as the closing bracket of the if branch in the console mode.

```
> day <- "Saturday"
> if (day == "Saturday" | day == "Sunday") {
+ print("It is the weekend")
+ print("It is good")
+ } else
+ {
+ print("It is a work day")
+ }
[1] "It is the weekend"
[1] "It is good"
```

When there are more than two alternative statements to be executed, we can simply use a nested conditional statement.

```
if (condition1) {
    # condition 1 is TRUE
    if (condition 2) {
        # condition 2 is TRUE
        ...
    } else {
        # condition 2 is FALSE
        ...
```

```
        }
    } else {
        # condition 1 is FALSE
        ...
    }
```

But what if we have many more alternatives to choose from? A structure with many nested conditional statements will became confusing and difficult to maintain. The `switch` statement can handle these situations effectively. It has the following syntax:

switch(*expression*, ...),

where ... represents a list of options. The `switch` statement accepts a string or number as the first argument (if the argument is not a string it is coerced to integer). When a number is the first argument the corresponding expression from the list of options will be evaluated and returned:

```
> day <- 2
> switch(day,"Mon","Tue","Wed","Thu","Fri","Sat","Sun")
[1] "Tue"
```

If the first argument is a string, an exact match with the names of the elements in the option list is searched. Then, the found element is evaluated and returned:

```
> day <- "Tue"
> switch(day,
+          Mon = "Monday",
+          Tue="Tuesday",
+          Wed="Wednedsay",
+          Thu="Thursday",
+          Fri="Friday",
+          Sat="Saturday",
+          Sun="Sunday")
[1] "Tuesday"
```

The `ifelse` function can be used for vectorized decisions. The returned value has the same shape as the first parameter. Its values are filled with values from the second or third parameter depending on whether the values of the first parameter are evaluated as TRUE or FALSE.

```
> x <- 1:5
> ifelse(x>3, 'greater than 3', 'less or equal to 3')
[1] "less or equal to 3" "less or equal to 3"
[3] "less or equal to 3" "greater than 3"
[5] "greater than 3"

> letters <- c("a", "b", "c", "d", "e")
> digits <- c(1, 2, 3, 4, 5)
> ifelse(c(TRUE, FALSE, TRUE, TRUE, FALSE), letters, digits)
[1] "a" "2" "c" "d" "5"
```

### 2.15.2 Loops

R provides three basic types of *loops*. The simplest loop, `repeat`, only repeatedly evaluates the given expression (infinite loop). The `break` keyword is used to terminate the loop.

```
> i <- 1
> repeat {print(i); i <- i+1; if (i>5) break}
[1] 1
[1] 2
[1] 3
[1] 4
[1] 5
```

The `while` loop implements a loop with the condition at the beginning. It repeats an expression as long as the loop condition is true.

```
> i <- 5
> while (i>0) {print(i); i <- i-1}
[1] 5
[1] 4
[1] 3
[1] 2
[1] 1
```

The `break` command can be used inside a while loop too. It terminates the loop here as well. In addition, the `next` statement skips the remaining statements in the loop body and starts a new loop iteration (evaluating the condition).

The for loop is used to iterate through each element of a vector or list. These elements are assigned to a variable, the name of which is a part of the for statement.

```
> a <- c(1, 2, 3)
> b <- c("a", "b", "c")
> for (i in a) print(i)
[1] 1
[1] 2
[1] 3
> for (i in 1:length(a)) print(a[i])
[1] 1
[1] 2
[1] 3
> for (i in c(a,b)) print(i)
[1] "1"
[1] "2"
[1] "3"
[1] "a"
[1] "b"
[1] "c"
```

The variable is, however, not an alias for the iterated values:

```
> a <- c(1, 2, 3)
> for (i in a) i <- i+1
> a
[1] 1 2 3
```

## 2.16 Packages

R offers many readily available functions. Sometimes, however, functions for other different tasks are needed too. The functionality other than that provided by the R core team is accessible through so called packages.

A *package* is set of functions, data, and help files that have been bundled together in the *library* subdirectory (the directory containing a package is called the library).

A package typically contains functions that are related to a specific task. They were written by the R community to solve problems in various disciplines. A core set of packages is already included in the installation of R. Other packages are stored and available in the official R repository known as *CRAN* (Comprehensive R Archive Network). CRAN is a collection of sites which carry identical material about R, including the R distribution, extensions, documentation, and binaries. The CRAN master site (*https://CRAN.R-project.org/*) is maintained at Wirtschaftsuniversität Wien, Austria and a complete list of mirrors can be found at *http://CRAN.R-project.org/mirrors.html*. CRAN includes approximately 13,000 additional packages and a list of the same can be found at *https://cran.r-project.org/web/packages/*.

To see the list of packages loaded by default, use the command `getOption` (`"defaultPackages"`). The result omits the `base` package, which is loaded every time.

In order to view the list of all packages installed by the user, the `installed.packages()` function can be called. Alternatively, the *Packages* pane in RStudio shows a table with this information.

To get the list of the packages currently loaded into R, we can use the function `.packages()`.

## 2.16.1   Installing Packages

To be able to use a package, it has to be downloaded and installed on a computer.

Use the `install.packages()` function to install a CRAN package in R. The parameter is a name of the package to be installed. This function downloads the package from a repository and installs it locally. In order to install two or more packages at the same time, a list of package names can be passed as the parameter.

In RStudio, it is possible to use the bottom-right pane *Packages* and the *Install button*. Be sure the *Install dependencies* option is checked.

An example of running a code where the necessary library is not installed can be found in Section 5.6.1.

Sometimes, packages are updated by their creators. To check which packages have been modified, we can use the `old.packages()` function. If a newer version is preferable for us, we can call `update.packages()`. This will update all packages interactively. If the argument `ask` is set to TRUE (default value), packages with a newer version are reported and for each one we can specify whether we want to update it. If we only want to update a single package, the best way to do it is to use `install.packages()` again. In RStudio, the *Update* button in the *Packages* pane can perform this task too.

To get some basic information about a package, use the `library()` function with the `help` parameter set to the name of the package, e.g., `library(help="tm")`.

## 2.16.2 Loading Packages

To load a package in R, we call the `library()` function where the parameter is the name of a package. We may also use the function `require()`. It is a similar function which accepts a slightly different set of arguments.

Functions from a package are available after loading the package. If more functions with the same name are in more than one package, the one that has been loaded last will be called. To call another one, a function name can be preceded by a package name and `::`.

To unload a package, we can call the `detach()` function.

## 2.17  Graphics

R includes three graphical systems. One of them is represented by the standard `graphics` package, the others use the `lattice` package and the `ggplot2` package. R includes tools for drawing most common types of charts like scatterplots, line charts, bar charts, pie charts, histograms, etc. as well as any complicated graphics. To create a graphical output, a series of functions are called. They create complete plots (high-level functions) or add some elements to existing plots (low-level functions). This approach follows the so called *painters model* where the graphics are created in steps and newer objects overlap the previous output [204]. The graphical output is printed on the screen (in one of the panes of RStudio) or can be saved in a wide variety of formats.

In this book, we will focus on the `graphics` package which includes the capabilities for some basic graphics while providing a wide variety of functions for plotting data. We present only some basics that you might need to draw and customize simple graphs. To see some basic information about the `graphics` package and its functions, call `library(help="graphics")`.

The most important function for creating graphics is `plot()`. It accepts objects as parameters and tries to plot them. In the simplest case, the arguments are values on the $x$ and $y$ axes. The graph is created according to the supplied parameters. To display all the argument names and corresponding default values of the `plot()` function, run the `args(graphics::plot.default)` statement.

To demonstrate some basic capabilities of the `plot()` function, we will use fictive data – numbers of newspaper articles from different categories (classes) published in five days of a week.

```
> articles <- matrix(c(3,10,6,8,15,9,2,9,4,6,8,3,3,7,1),
+                     nrow = 5,ncol = 3,byrow = TRUE)
> dimnames(articles) = list(
+     c("Mon", "Tue", "Wed", "Thu", "Fri"),  # row names
+     c("Economy", "Politics", "Science"))   # column names
```

```
> articles
      Economy Politics Science
Mon         3        10       6
Tue         8        15       9
Wed         2         9       4
Thu         6         8       3
Fri         3         7       1
```

To create a simple plot with the number of released articles in the *Economy* category on the *y* axis (see Figure 2.8), we can call `plot(articles[, 1])` or `plot(articles[, "Economy"])`.

We can change the way the values are displayed by providing a value for the type parameter, e.g., p for points, l for lines, b for both, o for both overplotted, h for histogram, or s for stair steps. The title is defined by the `main` argument or can be added by calling the `title()` function.

```
> plot(articles[,"Economy"], type = 'o')
> title(main="Articles numbers")
```

In the following example, we first display the plot with the number of articles from the *Economy* category. No axes are displayed (the axes parameter), they will be added later, the line color will be changed to blue (the col parameter), the y-axis will be in the range 0 to the highest value in articles matrix (the ylim parameter), and the axes will not be annotated (the ann parameter). Subsequently, we will add the main graph title, the x and y axes (at defines the points at which tick-marks are to be drawn, the labels parameter provides the labels), descriptions of the axes in dark grey color, and a box around the graph, see Figure 2.9).

```
> articles_range=range(articles)
> plot(articles[, "Economy"], type="o", col="blue",
+       axes=FALSE, ann=FALSE, ylim=articles_range)
> title(main="Numbers of articles")
> axis(1, at=1:5, labels=rownames(articles))
> axis(2, at=2*0:(articles_range[2]/2))
> box()
> title(xlab="Days", col.lab="darkgrey")
> title(ylab="Number of released articles",
+        col.lab="darkgrey")
```

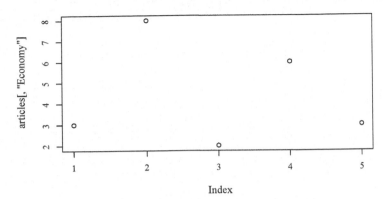

**Figure 2.8:** The simplest X-Y plot.

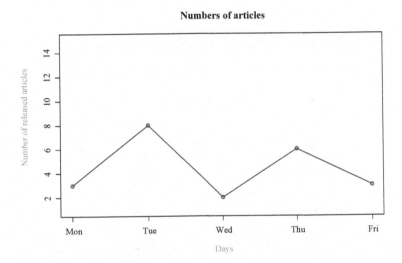

**Figure 2.9:** A customized X-Y plot.

We can add another line for the articles from the *Politics* category. It will be red in color, the line will be dashed, and the points will be square in shape. To clarify what is on the graph, a legend will be added. It will be placed on a specified position in the graph (coordinates [4, 15]), the text will be slightly smaller (the cex parameter), and the same line colors (the col parameter), point types (the pch parameter), and line types (the lty parameter) used by the actual plots will be used, see Figure 2.10).

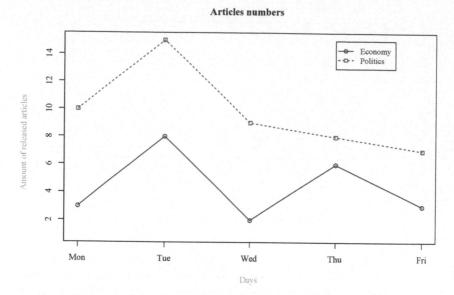

**Figure 2.10:** A customized X-Y plot with two data segments.

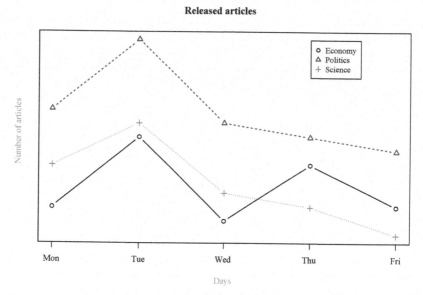

**Figure 2.11:** A customized X-Y plot with three data segments.

```
> lines(articles[,"Politics"], type="o", pch=22,
+        lty=2, col="red")
> legend(4, 15, c("Economy","Politics"), cex=0.9,
+        col=c("blue", "red"), pch=21:22, lty=1:2)
```

The plot() function is a good choice if you want to plot one or two columns in one chart. When more data segments are to be displayed in a plot, another function, matplot(), can be used instead (see Figure 2.11).

```
> matplot(articles, type ="b", pch=1:3, col=1:3,
+          axes=FALSE, ann=FALSE)
> title(main="Released articles")
> axis(1, at=1:5, lab=rownames(articles))
> axis(2, las=1, at=2*0:released_articles[2])
> box()
> title(xlab="Days", col.lab='darkgrey')
> title(ylab="Number of articles", col.lab='darkgrey')
> legend(4, 15, names(articles[1, ]), cex=0.9,
+          col=1:3, pch=1:3)
```

A pie chart is a good option to show some proportions of a whole, for example, when we want to show how are the newspaper articles in the *Economy* category are distributed in a week. This type of graph is created with the pie() function. The first parameter is a vector of non-negative numbers which are displayed as areas of pie slices. Other parameters include labels to be assigned to the slices, colors of the slices, or a title for the chart.

```
> colors <- c("green", "blue", "yellow", "orange", "red")
> pie(articles[, 1], main=names(articles[1, ])[1],
+      labels=names(articles[, 1]), col=colors)
```

Slices with the specified colors, labeled with days are displayed in the chart created. In the following example, we will add also the absolute numbers of released articles together with the percentages rounded to one decimal place, and a legend, see Figure 2.12).

```
> percent <- round(articles[, 1]/sum(articles[, 1]) * 100, 1)
> percent <- paste(percent, "%", sep="")
> pie(articles[, 1],
```

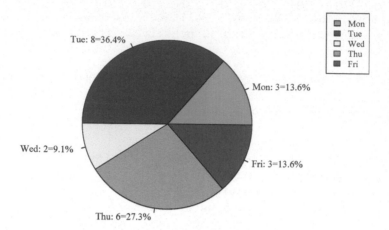

**Figure 2.12:** Pie chart with labels and a legend.

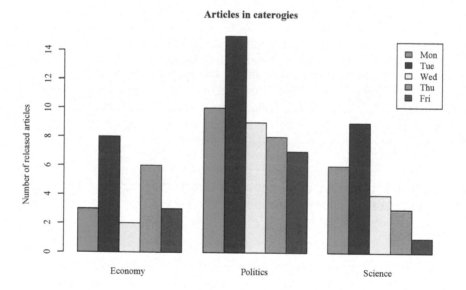

**Figure 2.13:** Bar chart.

```
+      main=names(articles[1, ])[1],
+      col=colors,
```

```
+          labels=paste(paste(names(articles[, 1]),
+                       articles[, 1],
+                       sep=": "),
+                       percent, sep="="))
> legend("topright", names(articles[, 1]),
+          cex=0.9, fill=colors)
```

A simple bar chart can be created using the `barplot()` function. If the first(or the `height`) parameter is a vector, a sequence of bars representing the numeric values of the vector will be created.

```
> barplot(articles[, 1])
```

If the `height` parameter is a matrix, the bars correspond to the columns of the matrix. The values in the columns are either stacked in one bar or represented by a sequence of other bars (depending on the logical `beside` parameter). In our example, the x-axis will represent the newspaper categories and the sub-bars the days of the week. The `t()` function can transpose the matrix so we can change the perspective on the data (i.e., days on the x-axis and categories as sub-bars).

```
> barplot(articles)
> barplot(articles, beside=TRUE)
> barplot(t(articles))
```

We can play with many parameters, for instance, adding descriptions of axes, changing colors, or adding a legend.

```
> barplot(articles, main='Articles in caterogies',
+          ylab='Number of released articles',
+          beside=TRUE, col=colors, legend=TRUE)
```

We can customize many features of the graphical output through setting graphical options using the `par()` function. This function, called without parameters, returns a list of the values of current parameters. To change the parameters, a list of `tag=value` pairs are submitted. The parameters include, for example, justification of texts, margins, font, foreground or background colors, and many others.

# Chapter 3

# Structured Text Representations

## 3.1 Introduction

*Structured* data is usually organized in databases, particularly relational databases. The high level organization, typically represented by tables, enables easy and efficient processing, for instance, searching or filtering. Text data is usually considered to be *unstructured*. A newspaper article, e-mail, SMS, or a recipe in a cookbook definitely does not look like a table. However, because the texts are normally generated using the grammar of a natural language and rules of a certain linguistic style, they have some kind of structure. For example, in a scientific paper, parts like a title, authors, abstract, keywords, introduction, etc. are expected or required. Every part consists of one or more words, symbols, expressions, sentences, organized into paragraphs, sections, or subsections. Every sentence contains words from several classes (e.g., nouns, verbs, adjectives) that are combined in the form of verb, noun, or prepositional phrases so they can bear some meaningful information. The words belong to a dictionary of the given natural language and their combinations are driven by the rules of the language. The rules are known as the *grammar*. There exist many rules (types of a sentence or expression) in a language and it is possible to create infinite texts written in a specific language. However, the structure is very complicated and it is not possible to describe all texts from one domain for example, the newspaper articles about the Winter Olympic Games in 2018 with one structure (template). An example of a structure would be all visitors of the Grandezza hotel in Brno, where every visitor has a name, surname, data of birth, address, etc. It is obvious that the second example data (hotel visitors) distinguishes some entities where all entities from the same class have the same attributes (properties). This is not true for the newspaper article text.

There exists another form of data between structured and unstructured data. It is the data that has some structure but this structure does not conform to what is normally expected from structured data. It means that this structure cannot be directly converted to a structure of a relational database (tables). This data is known as *semi-unstructured*. Textual documents, especially those on the web, are a typical representative of this. In the semi-structured form, the structure is usually expressed by using some tags or other marks which identify and separate pieces of data and thus, enable the creation of some data records and their relationships, often hierarchical (a tree of elements). The schema of semi-structured data does not have to be fixed in advance, which provides a high flexibility; hence, all instances of the same entity do not have to share the same attributes. Redundancy or non atomic values are also allowed [173, 219]. Typical representatives of semi-structured formats are XML or JSON.

It would be very complicated to process texts without some kind of simplification, i.e., giving them a better structure so they can be more easily understood by machines. In fact, most machine learning algorithms require the data to be processed in a highly structured format. A structured format, as we already mentioned, is usually something that can be represented by a table. In the rows, there are some objects to be processed. In the case of text documents, those objects might be e-mails to be classified as *spam* (unsolicited, junk e-mails) and *ham* (legitimate, relevant e-mails). In the columns of the table, there are features (properties, characteristics) of the objects and their specific values. Every object is thus represented by a *feature vector* that characterizes it.

The question is, how to derive the features from unstructured texts. In the case of e-mails (in the spam filtering problem), those features might be a result of tests performed by the SpamAssassin filter [89]. These tests include examinations whether an e-mail contains images, an HTML code, empty lines, words like *cialis*, *viagra*, or *Your Bills*, whether the sender is known or not, is in different blocklists etc. We might include other properties, such as the length of a document in words, the date of publication, author, number of figures and tables, language, and so on. In a majority of problems, however, the content of the documents needs to be taken into consideration. It means that the words, their combinations and meaning are of the highest importance. The document features are thus often derived from the words in the document.

A widely used format in the text mining domain is the *vector space model* proposed by Salton and McGill [236]. Every document is represented by a vector where the individual dimensions correspond to the features (terms) and the values are the weights (importance) of the features. All vectors then form so called a *document-term matrix* where the rows represent the documents and the columns correspond to the terms in the documents. The features often correspond to the words contained in the documents. In other words, the particular property of a document related to a specific word is determined by the occurrence of the word in the document.

Let's take the following group of documents (fictive news headlines):

■ *The Vegas Golden Knights ice hockey team achieved the milestone of most home wins.*

■ *Germany won 1-0 over Argentina in the FIFA Wold Cup finals.*

■ *Someone has won tonight's USD 10 million jackpot in the National Lottery.*

■ *A man has won a USD 5 million jackpot on a Las Vegas slot machine.*

From the words and expressions used in the sentences, one can conclude that the first two documents are about *sports* and the remaining two talk about *gambling*. How could we arrive at such a conclusion? In the first two documents, there are words like *ice*, *hockey*, and *FIFA* that are almost exclusively related to the sports domain. In the last two documents, words like *jackpot* or *lottery* are probably the most notable. We might therefore say that if a document contains the word *hockey* it is most likely related to sports. In other words, the document has the property of having the word *hockey* included in the text.

For the purpose of assigning a document to a particular category, the position of the word is sometimes not important. We might change the first sentence to *The ice hockey team of Vegas Golden Knights achieved the milestone of most home wins.* The word *hockey* moved to the third position in the sentence from the sixth position and the context (the surroundings, a few words before and after the word) also changed. However, the meaning remains the same and it is still a sentence related to sports.

It seems that treating the content of a document as a set of words appearing in it is sufficient in some situations, such as assigning document to a category. In such a set, any relations with the remaining elements are not relevant. These relations in a sentence represent the *context* of the words – which other words are in the same sentence, which words are before and which are after, at what distance they are, etc.

Not considering these relationships, however, definitely causes some information to be lost. Let us look at the word *Vegas*. Without considering its context, this word is related to both topics – *sports* and *gambling*. The relations to some topics are not clear also for some other words, like *ice*, *Golden*, *World*, or *machine*. Knowing their context, the situation might change significantly. If we have expressions like *ice hockey*, *Vegas Golden Knights*, *World Cup*, or *slot machine*, their meaning is much clearer than in the case of handling each word separately. The solution might be to not consider only isolated words but also sequences of two, three, or even more words (known as *n-grams*) or treating some word

```
IF document contains ''hockey''  THEN document category = sport

IF document contains ''lottery'' THEN document category = gambling

IF document contains ''Vegas'' AND
   document contains ''Knights'' THEN document category = sport

IF document contains ''machine'' AND
   document contains ''slot''    THEN document category = gambling

...
```

**Figure 3.1:** Examples of rules used for categorization of documents into two categories (sport and gambling).

sequences as *named entities* (e.g., *Vegas Golden Knights* is a name for an ice hockey team, *Las Vegas* is a name of a place).

The algorithms that are used to analyze texts often work properly with a combination of documents rather than just one document. The rules used to categorize the documents can take the form shown in Figure 3.1.

This means, that even when the words are treated independently, a state when their context is somehow taken into consideration can be achieved. For example, · the Naïve Bayes or decision tree learners handle the features independently. After training, a representation consisting of multiple attributes (the joint probability calculation and decision tree branches) is created (see Chapter 5). The independence of attributes is also sometimes necessary for making the problem feasible, as in the case of training the Naïve Bayes classifier. Imagine that we have 1,000 unique words ($n = 1000$) and we want to use them to create a document of 50 words ($r = 50$), where some of the words might appear more than once. Keeping in consideration the order of the words, there exist $n^r$ possibilities (permutations), which is $10^{150}$. When the order of words is not important, the number of possible documents (combinations) is $(r + n - 1)!/(r!(n - 1)!)$, which is about $1.097 * 10^{86}$. It is thus obvious that consideration of the word order significantly increases the complexity of the solved problems.

Documents are therefore often treated as sets of independent words or expressions. One word might appear multiple times in one document. The number of occurrences can also play an important role. The more times a word appears, the more important it probably is. Sets generally do not allow multiple occurrences of items. An exception is so called *multisets* or *bags* that allow multiple instances of their elements. In the text mining domain, the later term is commonly used and the *bag-of-words* model is a model commonly used in natural language processing. It become very popular because of its simplicity and the straightforward process of creation while providing satisfactory results [135].

## 3.2  The Bag-of-Words Model

In the bag-of-words model, the features that are derived from the processed documents are based on the words contained in them. Each document is thus represented by one bag (multiset). The words that appear in the document are present in this set. The words that are not present in the document are not in the set. The sets can be converted to vectors that are a more common data representation in machine learning. This approach is depicted in Figure 3.2. We can see that different documents are represented by sets of different sizes (containing different numbers of elements). The vectors created from the sets are also of different sizes. And what is more, we can see that the vector components (table headers) have different meanings. For example, the first component of the first vector represents the presence of the word *good* whereas the first component of the second vector represents the presence of the word *bad*. This is not a desired state when we want, for example, to compare these two documents. They have the same value of the first feature but the semantic meaning is opposite. Additionally, most of the machine learning algorithms require the vectors to have a fixed length.

These considerations lead us to the conclusion that, in the ultimate representation of individual documents to be processed, all documents must have the same number of features and we must be able to clearly identify each feature across the entire document collection. Thus, the attribute set for the collection

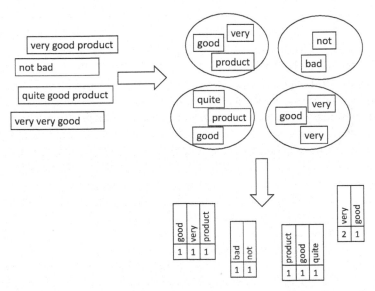

**Figure 3.2:** The principle of transforming documents (here, examples of very short product reviews) to bag-of-words.

| very | good | not | bad | quite | product |
|------|------|-----|-----|-------|---------|
| 1 | 1 | 0 | 0 | 0 | 1 |
| 0 | 0 | 1 | 1 | 0 | 0 |
| 0 | 1 | 0 | 0 | 1 | 1 |
| 2 | 1 | 0 | 0 | 0 | 0 |

**Figure 3.3:** A document term matrix created from the documents from Figure 3.2.

must be able to accommodate all possible features and all documents will share the same features. Of course, some of the documents may not be characterized by some features (i.e., their value is 0), see Figure 3.2. Now, all documents in a collection are represented by vectors of the same type and can be arranged into a table known as the *term-document matrix*. Such a matrix has a highly structured format which is suitable for many machine learning algorithms.

## 3.3 The Limitations of the Bag-of-Words Model

The application of the bag-of-words approach in creating the term-document matrix has a few disadvantages. First, there exist a large number of possible features which can be derived from texts. When considering words as the most natural features, even quite a small collection of documents can contain many unique words. In a data set consisting of 100 hotel accommodation reviews from *booking.com* where the average number of words in a review is about 25, almost 1,000 unique words can be found. In 1,000 documents, we identified more than 3,500 unique words, a collection having 10,000 documents contained almost 13,000 unique words, and in 100,000 documents, about 45,000 distinct words could be distinguished [295]. Such a high number of features is significantly different from what is typical in classic data mining problems or in relational databases.

The high number of features is often a problem from the computational complexity perspective because processing a larger number of attributes requires more time. For some algorithms, for example a decision tree induction, the computational complexity grows quadratically with the number of features [144].

Another problem related to the high number of dimensions is known as the *curse of dimensionality*. Inductive machine learning algorithms try to find a generalization of the available data. In the case of supervised classification, the goal is to find a function that is able to assign every item to a correct category (class). Having a data set with n binary (or boolean) attributes, there exist $2^n$ different instances. When there are 100 binary attributes, there exist $1.27 \cdot 10^{30}$ instances. To learn a perfect function covering the entire instance space, we would need a huge number of these instances. This is, however, often neither possible nor feasible. Because of that, we expect that the classifiers will work with some errors and not so many learning instances are needed. With the use of the PAC-learning (probably approximately correct learning) paradigm, we can even quantify the relationship between these numbers [150].

The curse of dimensionality is a bigger problem for some algorithms that for others. For example, a k-nearest neighbors (k-NN) classifier uses all features to calculate a similarity between the items in a data set. Thus, the k-NN classifier is more sensitive to the high number of attributes than, for example, a decision tree or rule learners that work with only a subset of attributes when creating a generalization of the data [199].

There are two famous statistical laws applicable to natural languages – the Zipf's law and Heaps' law *Zipf's law* says that the frequency of a word (or another linguistic unit, like a word stem) is inversely proportional to its rank derived from the frequency of the word in the collection of texts [186]. This relationship can be represented by the following equation:

$$freq(w_i) = \frac{C}{i^a},$$
(3.1)

where $freq(w_i)$ is the number of occurrences of the word $w_i$ and $i$ is the position of the word $w_i$ in the list of all words ordered in descending order according to their frequency (in other words, $i$ is the rank of the word $w_i$). The coefficients $C$ and $a$ can be derived from the available data having the word counts for every word appearing in the document collection. Ideally, $C = freq(w_1)$ and $a = 1$. In that case, the frequency of the second most frequent word will be one half of the frequency of the most frequent word, the frequency of the third most frequent word will be one third of the frequency of the most frequent word, etc. The frequency of the $10^{th}$ most frequent word will be twice as much as the frequency of the $20^{th}$ most frequent word and three times the frequency of the $30^{th}$ most frequent word. Zipf's law was slightly modified by Mandelbrot [184] to better represent the law, especially for words with high and low ranks:

$$freq(w_i) = \frac{P}{(i+\rho)^B},$$
(3.2)

where $freq(w_i)$ is the number of occurrences of the word $w_i$, $i$ is the rank of the word $w_i$, and the coefficients $P$, $\rho$, and $B$ are derived from the data.

*Heaps' law* [114] describes the relationship between the size of a data collection as expressed by the number of words and the number of unique words. This relationship is demonstrated by the formula:

$$M = k \cdot N^b,$$
(3.3)

where $M$ is the number of unique words, $N$ is the total number of words, and $k$ and $b$ are coefficients derived from data. Manning, Prabhakar and Schütze [185] derived the values $k = 44$ and $b = 0.49$ for the Reuters-RCV1 data set. The typical values of $k$ are between 30 and 100, $b$ is close to 0.5 (the number of unique words is thus proportional to the square root of the total number of words).

The coefficients that appeared in the formulae representing the laws of natural languages often depend on a specific language; for example, how inflective it is [98]. However, the laws can give us at least some initial insight to the data and its statistical characteristics.

For example, when considering the first 10 most frequent words, they should form about one third of all words in a document collection keeping in mind the ideal set that conforms to Zipf's law. Since the data is not ideal, this is not very exact. However, Manning, Prabhakar and Schütze [185] found that the 30 most common words account for about 30% of the tokens in Reuters-RCV1. Not using these words as features in a machine learning task will, on the one hand, decrease the number of features by a negligible amount. On the other hand, it will significantly decrease the number of non-zero attribute values. Looking at the ranked list of words from another perspective, a significant majority of words appear only a few times. These words are thus often irrelevant for many tasks (e.g., classification, clustering, or association mining) and can be ignored. They also often represent some spelling errors so they actually work as a noise. By removing these words from further processing, a substantial reduction of the number of features can be achieved.

The large number of features and the implications of Zipf's law also lead to an additional problem. When the number of words in the documents is considered, it quite small compared to the number of all unique words in a document collection. Looking at the previously mentioned collection of customer reviews, we were able to identify almost 1,000 unique words for the small set consisting of 100 documents. Taking the average length of 25 words, about 2.5% of the vector elements have a non-zero value (when all words in a document are distinct). In the case of the largest collection of 100,000 documents, the percentage of non-zero items is much smaller, about 0.05%. We therefore say that the vectors are very *sparse* which is a problem for some tasks, computing similarity in classification or clustering [177], for example. The sparsity can also be seen in the columns of the document-term matrix because a significant number of words appear only a few times. The sparsity generally means that it is, at times, quite complicated to identify some patterns in the data [87].

Since word order is not considered, the bag-of-words model has several disadvantages. For example, two sentences with the same words that appear in a different order will be treated as identical. When word order in a comparative sentiment analysis task is not taken into consideration, the sentences *A is better than B* and *B is better than A* will look the same although they should lead to opposite conclusions.

The representations of individual words also bear little semantics. Each word in the bag-of-words model is just one column in the document-term matrix where the order of columns is not important. When columns A and B are closer to each other than columns A and C, it does not mean that the words A and B are more similar in their meaning or grammatical form, for example, than the words A and C. In the vector space model, all words create one dimension that is equally important as the others. The distance from the beginning of the vector space of a position representing, for example, one occurrence of the word A is the same as the distances of other positions representing one occurrence of any other word.

The issue of the large number of features can be solved by feature selection methods (see Chapter 14). The bag-of-words model can be also used to derive new representations (using, for instance, latent semantic analysis or neural models creating word embeddings) with fewer dimensions, thus eliminating the above mentioned problems. These methods are briefly described in Chapter 13. The loss of context can be eliminated by working with sequences of words known as n-grams (we therefore sometimes talk about *bag-of-n-grams*) or by introducing new attributes bearing information about the context of a word (for example, what is the preceding word, what is the following word, what is the part-of-speech of the previous word, etc.). Here, however, problems of high dimensionality and sparseness are often intensified.

Despite the numerous problems and limitations of the bag-of-words model, it brings surprisingly good results in many tasks. In situations, especially when there are fewer features, where there is no correlation between them or when we do not want to take it into account, the bag-of-words approach is appropriate. Sparse vectors can be also used to learn dense embedding through training a neural network [101]. For example, combining the embedding vectors of words (through averaging them, for instance) from a document into one vector enables application of classical machine learning algorithms for classification.

## 3.4 Document Features

The form taken by the features characterizing documents depends on the problem to be solved. For example, using characters as features might be somehow useful for a spam filtering task where the presence of some special characters or too many identical characters can indicate a spam e-mail. On the other hand, in an information extraction problem where some kind of semantic information understandable by people is expected as the output, characters as features do not make much sense. Here, words or concepts play a more important role.

In most cases, the following feature types can be distinguished [87]:

■ *Characters, character bi-grams, tri-grams, n-grams:* In spite of being the most complete representation of textual data, characters are not often used as features. The reason is quite simple – one character does not bear much semantic information, so any interpretation is quite complicated, especially when the positional information, i.e., information about the context of a character, is lost (we can thus talk about a bag-of-characters representation).

■ *Words:* Words are the most natural and straightforward features that can be derived from texts. A *word* is a sequence of symbols, typically letters, that is used in the sentences of a given language. It has a semantic, grammatical, and phonological unity and is usually separated from other words by spaces in writing [74]. This is how are words understood by most western languages. On the other hand, texts in, for instance, Chinese or Arabic do not usually contain spaces between words so it is not easy to visually identify their borders.

A *dictionary* can be used to find out what is and what is not a word in a given text. It is quite typical that a dictionary, a list of words that are available in the given language, exists for a natural language. The dictionaries might contain only the basic form of the given word, called *lemma*. For example, the lemma for *does, doing, done*, or *did* is *do*. Therefore, the absence of a particular sequence of characters in a dictionary does not necessarily mean that it is not a proper word.

What about other parts of texts, like numbers, temporal expressions, currency symbols, and punctuation (comma, period, question mark, quotation marks, etc.)? It generally depends on the task to be performed. Sometimes, a particular number or date can be important. If the number is, for example, the price of a product, we can use a specific value to find out whether the product is expensive or cheap. Sometimes it is enough to know that the text shows specific price.

One of the disadvantages of using words is that, without a context, the possible meaning of a word may not be obvious.

■ *Terms:* Terms are single words or multiword expressions formed of words from the text. When looking at the fictive news headlines mentioned earlier in this chapter, terms like *Las Vegas, ice hockey, FIFA World Cup*, or *the National Lottery* can be identified. Since more words are treated as one unit, the ambiguity of words is reduced.

Term extraction is related to the *Named Entity Recognition* (NER) problem, which is a subtask of *Information Extraction*. Its goal is to identify information units like names (typically persons, organizations, and location) and numeric expressions including time, date, money, and percent expressions [205].

■ *Concepts:* Concepts are features that are generated from the text based on the occurrence of words, phrases, or more complicated structures. The concepts do not have to appear directly in the text. In the case of news headlines, for example, concepts like *sport* or *gambling* can be identified.

Concepts, together with terms, have the most expressive semantic value. Concepts enable better handling of synonymy, polysemy, hyponyms, or hypernyms. On the other hand, they require more effort to be extracted. They are also often domain dependent.

Assigning correct concepts often involves cross-referencing against a source of external knowledge which can be the knowledge of a domain expert, an existing ontology or lexicon, or an annotated collection of training documents used to train an algorithm assigning concepts.

Characters or words can be combined to form character or word n-grams. A character n-gram is a sequence of n consecutive characters. A few character 3-grams created from this sentence include "A f", " fe", "few", "ew ", "w 3", etc. Word 2-grams from this sentence include "Word 2-grams", "2-grams from", "from this", etc.

The character n-grams are able to capture whole words ("few") as well as word class (e.g., "ed ", "ing"). A great advantage of character n-grams when compared to word n-grams is that they do not create as many dimensions (the sparsity of data is lower) because there exist fewer character combinations than word combinations [139]. Character n-grams have been successfully used for many tasks, including language identification [38], authorship identification [118], malicious code detection [223], and plagiarism detection [17].

Word n-grams can help in capturing multiword phrases [163] and have been successfully used in sentiment analysis [273].

An interesting collection of word n-grams, together with their frequency of usage in books, is provided by Google. Their new edition of the *Google Books Ngram Corpus* provides syntactically annotated n-grams for eight languages and covers about 8% of books that have been ever published (more than 8 million books) [172].

N-grams can be also understood as a simple statistical language model that models probabilities of sequences of characters or words. In other words, given a sequence of previous words, the model can help us predict a word. As such, it can be successfully used in task, such as machine translation, spelling correction, or speech recognition [138].

## 3.5 Standardization

Text data is commonly presented in some kind of semi-structured representation. A web page containing the most recent news might include several news features consisting of a heading, list of keywords, a few paragraphs of text, images with captions, etc., all enclosed within <div> or similar tags with defined classes.

For example, an XML file containing news from Reuters Corpus Volume 1 (RCV1) [167] consists of elements containing the publication date, related topic, mentioned people, places or organizations and the text consisting of a title, date line, and body (see Figure 3.4).

```
<?xml version="1.0" encoding="iso-8859-1" ?>
<newsitem itemid="2286" id="root" date="1996-08-20" xml:lang="en">
  <title>MEXICO: Recovery excitement brings Mexican markets
          to life.
  </title>
  <headline>Recovery excitement brings Mexican markets
            to life.
  </headline>
  <byline>Henry Tricks</byline>
  <dateline>MEXICO CITY</dateline>
  <text>
    <p>Emerging evidence that Mexico's economy was back on the
       recovery track sent Mexican markets into a buzz of excitement
       Tuesday, with stocks closing at record highs and interest
       rates at 19-month lows.
    </p>
    <p>&quot;Mexico has been trying to stage
    ...
    </p>
  </text>
  <copyright>(c) Reuters Limited 1996</copyright>
  <metadata>
    <codes class="bip:countries:1.0">
      <code code="MEX">
        <editdetail attribution="Reuters BIP Coding Group"
                    action="confirmed" date="1996-08-20"/>
      </code>
    </codes>
    <codes class="bip:topics:1.0">
      <code code="E11">
        <editdetail attribution="Reuters BIP Coding Group"
                    action="confirmed" date="1996-08-20"/>
      </code>
      ...
    </codes>
    <dc element="dc.publisher" value="Reuters Holdings Plc"/>
    <dc element="dc.date.published" value="1996-08-20"/>
    ...
  </metadata>
</newsitem>
```

**Figure 3.4:** An example of an XML file containing a newspaper article from the RCV1 document collection.

A JSON (JavaScript Object Notation) file stores the data in a human readable format as objects consisting of name/value pairs where the value can be a scalar (string, number, logical value), another object, or a list of values [79].

With a semi-structured document, only the content of some selected elements might be important for a specific task. For example, to train a classifier to categorize newspaper articles, only the content of elements <code> in the <codes class="bip:topics:1.0"> section, <title>, and <text> is relevant. The content of <title> and <text> will be the base for deriving document features, the content of <code> element will be used to specify the class labels (one or more) for the documents.

The XML library can be used to process XML (and HTML) documents. The xmlToList() function converts an XML document of node to a list that directly contains the data. The list elements correspond to the children of the top-level node and the .attrs element is a character vector of attributes from the node.

```
> library("XML")
> data<-xmlToList("Reuters/CD1/2286newsML.xml")
> names(data)
[1] "title"     "headline"  "byline"    "dateline"  "text"
[6] "copyright" "metadata"  ".attrs"
> data["title"]
$title
[1] "MEXICO: Recovery excitement brings Mexican
       markets to life."
> data$metadata$codes$code$.attrs
  code
"MEX"
```

The following example (from [143]) shows how to extract information about reviews of conference papers. The file with the information in the JSON format contains one object with key papers. The value associated to this key is an array where every element represents one paper. Every paper has a numeric identifier, a string value representing a preliminary decision on the paper, and the list of reviews. The reviews contain information about reviewers' confidence, evaluation of the paper on a given numeric scale, the number of the review, language, text of the review, additional remarks to the conference committee, date of the reviews, and authors' evaluation of the review. An excerpt from the JSON file can be found in Figure 3.5.

The fromJSON() function from the rjson package converts a string containing a JSON object, the content of a file, or the content of a document with a given URL to an R object. The type of the object is list. Its elements are num-

```
{
  "paper": [
    ...
    {
      "id": 8,
      "preliminary_decision": "accept",
      "review": [
          {
            "confidence": "4",
            "evaluation": "1",
            "id": 1,
            "lan": "en",
            "orientation": "0",
            "remarks": "",
            "text": "The paper describes an experience concerning...",
            "timespan": "2010-07-05"
          },
          {
            "confidence": "4",
            "evaluation": "2",
            "id": 2,
            "lan": "en",
            "orientation": "1",
            "remarks": "",
            "text": "This manuscript addresses an interesting solution...",
            "timespan": "2010-07-05"
          },
    },
    {
      "id": 12,
      "preliminary_decision": "accept",
      "review": [
          {
            "confidence": "4",
            "evaluation": "2",
            "id": 1,
            "lan": "es",
            "orientation": "2",
            "remarks": "Se trata de un articulo que presenta un tema ...",
            "text": "Es un articulo que presenta un tema novedoso ...",
            "timespan": "2010-07-05"
          },
          ...
    },
    ...
  ]
}
```

**Figure 3.5:** An example of a JSON file containing information about reviews of papers from a conference.

bered when they come from a list of values in a JSON object and have names when they are converted from an object with name/value pairs.

```
> library("rjson")
> data <- fromJSON(file="reviews.json")
> names(data)
[1] "paper"

> names(data[['paper']][[1]])
[1] "id"                   "preliminary_decision"
[3] "review"

> names(data[['paper']][[1]][["review" ]][[1]])
[1] "confidence"  "evaluation"  "id"         "lan"
[5] "orientation" "remarks"     "text"       "timespan"

> # a piece of text of the first review of the first paper
> substr(data[["paper"]][[1]][["review"]][[1]][["text"]],
+          1, 60)
[1] "- El articulo aborda un problema contingente y muy
relevante"

> # the language of the first review of the first paper
> data[["paper"]][[1]][["review"]][[1]][["lan"]]
[1] "es"

> # a decision on the first paper
> # same as data[[1]][[1]][[2]]
> data[["paper"]][[1]][["preliminary_decision"]]
[1] "accept"
```

The following commands convert the data to a data frame with two columns, Text and Evaluation. The data can be used, for example, in a classification task where the evaluation will be predicted according to the text of reviews.

```
# creating an empty data frame
d <- data.frame("Text"=c(),
                "Evaluation"=c(),
                stringsAsFactors=FALSE)
for (i in 1:length(data[["paper"]])) {
  # looking at the number of reviews of the ith paper
```

```
nr <- length(data[["paper"]][[i]][["review"]])
if (nr==0) {next}
for (r in 1:nr) {
  # adding the text and evaluation to the data frame d
  d <- rbind(d,
                data.frame(
                  "Text"=data[["paper"]][[i]][["review"]]
                            [[r]][["text"]],
                  "Evaluation"=data[["paper"]][[i]]
                                  [["review"]][[r]]
                                  [["evaluation"]]
                )
              )
}
}
```

## 3.6   Texts in Different Encodings

Texts are stored in computer memory as a sequence of bits. Each character then usually occupies a certain number of bytes. To order to be stored, a character needs to be encoded to a number. It needs to be decoded to display a character.

In the past, seven bits enabling the storage of 128 distinct characters were needed to encode a character. This was enough for English letters, digits, and some other common symbols like punctuation.

However, for some other languages, like German, Spanish, French, Czech, etc., characters with accents were needed. It was not possible to encode all these characters to a 7-bit character, so eight bits were used. Different languages had different encodings. For some languages, like Chinese, even eight bits were not enough [210]. This problem can be solved using the Unicode encoding, which is a de facto standard in simultaeously encoding many languages and scripts [109].

Knowledge of text encoding is necessary for some of the preprocessing steps, like tokenization, for example. In some languages, characters encoded by some numbers are regular word characters while in others they are punctuation characters [210].

Document encoding can be explicitly indicated as, for example, in some metadata or headers of a document. In certain contexts, an explicit or implicit agreement on a specific encoding might be made as well. When no such information is available, a user or computer must make a guess based on an inspection of the text [149].

To guess a document encoding in R, the `stri_enc_detect()` function from package indexR packages!stringistringi can be used. The function accepts a character vector and returns a list of guessed encodings, language, and confidence level for each string. It currently supports about thirty different character sets and languages.

```
texts <- c("What is your name?",
           "¿Cómo te llamas?",
           "Comment tu t'apelles?")
stri_enc_detect(texts)
[[1]]
      Encoding Language Confidence
1  ISO-8859-1       en       0.47
2  ISO-8859-2       hu       0.31
3       UTF-8                0.15
...
[[2]]
      Encoding Language Confidence
1       UTF-8                0.80
2  ISO-8859-1       es       0.47
3  ISO-8859-2       hu       0.15
...
[[3]]
      Encoding Language Confidence
1  ISO-8859-1       fr       0.71
2       UTF-8                0.15
3  ISO-8859-2       hu       0.14
```

Texts can be converted from one encoding to another using the standard `iconv()` function. The list of available encodings is provided by the `iconvlist()` function.

```
> x<- "Gr\xfc\xdf  Gott"
> x
[1] "Grüß Gott"
> Encoding(x)
[1] "unknown"
> Encoding(x) <- "latin1"
> Encoding(x)
[1] "latin1"
> xx <- iconv(x, from="latin1", to="UTF-8")
> Encoding(xx)
[1] "UTF-8"
```

## 3.7 Language Identification

Some of the preprocessing steps, like tokenization, stemming, or part of speech tagging are language dependent. Language identification is actually a categorization problem – a language label is assigned to a text.

A straightforward approach can compare words of a text with dictionaries of known languages and assign the language with the highest match as the correct one. Typically, texts are compared with only the most frequent words in a language. The accuracy of these methods is, however, not very high, especially for short texts [106].

Other approaches work with variable or fixed length character n-grams. The n-grams in a text are compared to an n-grams based model of a language. The model can be based on simple frequencies of n-gram occurrences [44] or can have the form of a probabilistic Markov model [215].

A comparison of some popular language identification algorithms can be found in [131].

To determine a language, for example, R provides, the cld2 (Google's Compact Language Detector 2) package and its function detect_language().

```
texts <- c("What is your name?",
           "¿Cómo te llamas?",
           "Comment tu t'apelles?",
           "Wie heißt du?",
           "Jaké je tvoje jméno?")
detect_language(texts)
[1] "en" "es" "fr" "de" "cs"
```

## 3.8 Tokenization

Tokenization is the process of splitting a document into pieces of texts known as *tokens*. These tokens are often the words contained in the text. In most European languages where the words are space delimited, the task seems to be quite simple – to split the text at the places where are white spaces. In some other languages, like Chinese, where there are no spaces between words, the text needs to be analyzed in greater depth.

Even in European languages, the process is usually not that straightforward. Punctuation (including brackets, apostrophes, dashes, quotation marks, period, exclamation mark) is usually understood as something separating the tokens. Besides removing white spaces, punctuation marks are usually removed too which is probably the simplest tokenization approach [275, 211].

There are, however, many exceptions and special problems related to different languages. Even for English, tokenization is not an easy task. The following text mentions a few typical problems.

A period is usually understood as the end of a sentence. It can be, however, used for other purposes, like in abbreviations (*Mr.*, *etc.*, *U.S.A.*) or in numbers in English (*1.23*). In such cases, the period is often a part of the tokens.

Apostrophes can be used in contractions (*I'm*, *haven't*), genitive forms of nouns (*Peter's*), or when omitting first two digits in a year (*'90s*). We then need to decide whether the apostrophe should be removed (*haven't* → *haven* + *t*), retained, or replaced (*'90s* → *1990s*). In many situations, it is still not clear what should happen. For example, should *doesn't* be transformed to *doesnt*, *doesn+t*, or *does+n't*?

Several words can be also connected by hyphens to create compound words (*state-of-the-art*, *bag-of-words*, *two-year-old*, *sugar-free*, *long-lasting*). Should we treat each part of the compound expression separately or the whole expression as one token?

Special attention should be given to numbers. Numbers can be written in many different ways, including a decimal point (comma in some languages), + or − sign, semilogarithmic notation, etc. When a phone number is treated as one token we cannot see, for example, an area code (*00420-777777777* vs. *00420+777777777*). The same might be applied to dates, *2018-02-14* can be treated as one token as well as *2018+02+14* so we have information about a year, month, and day.

Other pieces of text that require special attention are, for example, e-mail addresses, URLs, special names (*C++*, *Bell&Ross*) among others.

Some languages, like German, often use compound words that are not hyphenated. For example, the longest German word in 2013 [265] *Rindfleischetiketierungsüberwachungsaufgabenübertragungsgesetz* refering to the *law for the delegation of monitoring beef labelling* consists of ten words. Some of them are prefixes, but can also be prepositions and are counted as words here. Thus, the long word consists of ten words.

It is obvious that tokenization is a language specific task. The language of a text must therefore be known for tokenization that is not naive. The rules defining some specific tokens in a language must be clearly stated. The process of tokenization is also influenced by the used text encoding because delimiters and special symbols occupy different positions in different character tables [211].

## 3.9 Sentence Detection

In some tasks, splitting a document into smaller units, like sentences, can be beneficial. It seems that finding symbols which normally end sentences, like ., ?, or ! should be enough. These symbols can, however, also be parts of other expressions (for example, a period can be a part of an abbreviation or a number). A terminating symbol can be also omitted, like in the case when a sentence ends with an abbreviation.

A human expert can define a set of rules to be used to separate sentences (e.g., the symbol, like a period, must be followed by a whitespace and an upper-case letter). As manual creation of such rules need not be the best approach, an algorithm can be trained to detect sentences from labeled data [225].

In R, sentences can be extracted from a text using the udpipe() function from package udpipe. For more details, see the example at the end of this chapter.

```
> unique(udpipe("I'm Mr. Brown. I live in the U.S.A.",
+              object="english"
+              )['sentence']
+       )
             sentence
1        I'm Mr. Brown.
6 I live in the U.S.A.
```

## 3.10  Filtering Stop Words, Common, and Rare Terms

Some of the words that are too common do not usually contribute (or contribute only negligibly) to achieving a certain goal. For example, in the query *Find a hotel in Prague* submitted to a web search engine, the word *a* does not influence what will be retrieved by the search engine. The retrieved results will be the same no matter whether *a* is or is not in the query. The reason is simple – almost all documents written in English contain at least one occurrence of the word *a*. On the other hand, the word *Prague* is very important because without it the search engine would return information about hotels in many other locations.

Similar importance of words is apparent in other text mining tasks, such as classification or clustering. When categorizing, for example, newspaper articles to categories, the occurrence of word *the* will not play an important role because articles from all categories are very likely to contain this word. In a clustering process, when calculating the similarity between two documents, the occurrence of the word *the* will contribute to the similarity measure with the same value for almost all document pairs.

It is obvious that the words not contributing to achieving a certain goal are useless and do not have to be considered in further processing.

The words that are not important for a certain task are known as *stop words*. Very often, they are the most frequent words in a given language. Typical stop lists for English contain words like *the, of, and, by, with* (grammatical articles, prepositions, conjunctions, auxiliary verbs, pronouns, articles). Many stop word lists can be found on the web or incorporated in various tools and in the libraries of programming languages. Table 3.1 shows how the numbers of stop words for different languages can differ (the table is based on stop lists in python's NLTK library v. 3.4). It is, however, obvious that the number of stop words is usually around a few hundred. It should be noted that the available stop word lists sometimes suffer from surprising omissions and inclusions, or are incompatible with particular tokenizers [207]. Various stop word lists in R can be obtained from the function `stopwords()` from the `tm` package.

The stop word list length, when compared to the size of the dictionary, which can be a few hundreds of thousands, is negligible. The number of features thus does not decrease significantly when stop words are removed. A massive reduction is, however, noticeable in the total number of words (which is related, for example, to the memory requirements). According to Zipf's law, ten most frequent words represent approximately one third of all words in a corpus.

A stop list can also be easily created by simply counting the frequencies of words in a corpus and looking at the most frequent words. It is, however, better to carefully examine the frequent words so that important words are not lost.

What is and what is not a stop word might depend on a domain. In a collection of hotel accommodation reviews, the word *hotel* would probably be very frequent and will not be useful in, for example, determining whether a review is negative, positive, or neutral. In the following example, reviews from *booking.com* (for a detailed description of the data, see [295]) are processed. The number of all words, unique words, and global words frequencies are calculated from a matrix representing the documents. The frequencies together, with relative frequencies of the ten most frequent words are printed. Subsequently, the 50 most frequent words are displayed without their frequencies. We can see that the list contains typical English stop words as well words typical for the hotel accommodation domain. From a sorted list of words together with their frequencies, a graph demonstrating Zipf's law can be easily generated (with logarithmic axes).

**Table 3.1:** Numbers of stop words for different languages in stop lists in Python's NLTK library v. 3.4.

| Language | Number of Stop Words |
|----------|----------------------|
| English  | 179 |
| German   | 231 |
| Italian  | 279 |
| French   | 155 |
| Spanish  | 313 |
| Russian  | 151 |

```
> library(tm)

> text <- readLines("reviews-booking.txt")
> corpus <- VCorpus(VectorSource(text))
> corpus <- tm_map(corpus, removePunctuation)
> corpus <- tm_map(corpus, stripWhitespace)
> corpus <- tm_map(corpus, content_transformer(tolower))
> dtm <- DocumentTermMatrix(corpus)

> mat <- as.matrix(dtm)

> word_frequency <- colSums(mat)
> number_of_words <- sum(word_frequency)
> unique_words <- length(word_frequency)

> paste("The number of all words: ", number_of_words)
[1] "The number of all words:  394993"
> paste("The number of unique words: ", unique_words)
[1] "The number of unique words:  19528"

# printing a word together with its rank, total frequency,
# and relative frequency
> words_sorted <- sort(word_frequency, decreasing = TRUE)
> for (i in 1 : 10) {
+    print(paste("word: ", i, ": ",
+                names(words_sorted[i]),
+                ", freq.: ", words_sorted[i],
+                ", rel. freq.: ",
+                round((words_sorted[i]/number_of_words),3),
+                sep=""
+               )
+         )
+ }
[1] "word 1: the, freq.: 33313, rel. freq.: 0.084"
[1] "word 2: and, freq.: 18681, rel. freq.: 0.047"
[1] "word 3: was, freq.: 13445, rel. freq.: 0.034"
[1] "word 4: very, freq.: 8644, rel. freq.: 0.022"
[1] "word 5: room, freq.: 6969, rel. freq.: 0.018"
[1] "word 6: hotel, freq.: 6742, rel. freq.: 0.017"
[1] "word 7: for, freq.: 6092, rel. freq.: 0.015"
[1] "word 8: staff, freq.: 5447, rel. freq.: 0.014"
[1] "word 9: good, freq.: 4819, rel. freq.: 0.012"
[1] "word 10: not, freq.: 4361, rel. freq.: 0.011"
```

```
# printing 50 most frequent words
> names(words_sorted[1:50])
 [1] "the"          "and"         "was"      "very"
 [5] "room"         "hotel"       "for"      "staff"
 [9] "good"         "not"         "were"     "location"
[13] "breakfast"    "with"        "but"      "clean"
[17] "friendly"     "from"        "nice"     "had"
[21] "there"        "helpful"     "rooms"    "that"
[25] "have"         "excellent"   "you"      "this"
[29] "great"        "our"         "are"      "all"
[33] "comfortable"  "stay"        "would"    "they"
[37] "small"        "close"       "only"     "one"
[41] "well"         "could"       "which"    "food"
[45] "service"      "city"        "also"     "night"
[49] "bathroom"     "when"

# plotting the curve demonstrating the Zipf's law
> plot(words_sorted, type="l", log="xy",
+       xlab="Rank", ylab="Frequency")
```

Special attention should be paid to stop words in multiword expressions and to words that are somehow affected by the occurrence of a stop word. Consider the following phrases: *room with balcony* and *room without balcony*, or *good* and *not good*. Leaving the stop words (*with*, *without*, *not*) out of consideration, the meaning of the phrase will change significantly (in fact, it will be the opposite). There are also phrases consisting solely of stop words, like *The One* (a movie), *to be or not to be* (a famous phrase from Shakespeare's Hamlet), or *The Who* (an English rock band). Weiss et al. [275] thus recommend the removal of stop words after the features for a document set are created and other preprocessing techniques (e.g., stemming) have been applied.

While stop words appear at the top of a list of words ordered by their frequency, the words at the other end of the list can also be eliminated. Consider a classification problem where there are 1,000 documents in two classes, $C_1$ and $C_2$. Only one document, $d$ from the entire collection contains one specific word $w$. The document belongs to class $A$. The word $w$ is definitely helpful and can be used when assigning documents to $A$. The rule *if a document contains w it should be assigned to A* will cover 1/1000 of the training examples and is therefore not too general. Having too many such rules will very likely lead to overfitting of a classifier which is not desirable. Not considering $w$ by a classifier can cause 1/1000 = 0.1% error on the training data which is also negligible.

In the calculation of the similarity of two documents during clustering, the words with low frequencies will not have an effect on the results because there

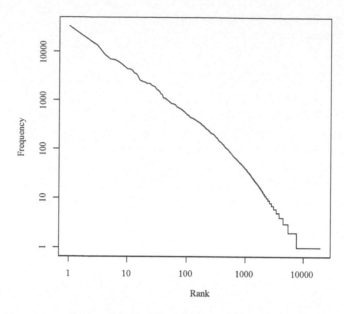

**Figure 3.6:** Graphical demonstration of Zipf's law.

is actually no similarity in the given dimension when one document contains a word and the other does not.

It seems that removing rare words is a good idea for most of the tasks. Sometimes, the computational complexity of a problem can be lowered significantly, especially for algorithms, where the complexity grows exponentially with the number of features. On the other hand, in an information retrieval task, not considering those rare words (for example, a specific name of a person, place, or company) can result in the failure to achieve a certain goal. Here, the document containing one specific word will not be found by a search engine. Indexing rare terms thus increases the success rate of a search engine but, on the other hand, might decrease its performance because more terms would need to be stored and processed.

## 3.11 Removing Diacritics

Diacritics (diacritical marks, accents) are symbols added to letters, usually above or below. Diacritics often changes the way a sound is pronounced (e.g., longer, with an accent). In some languages, like in Vietnamese, Czech, Spanish, or French, diacritics are quite common. In some situations, they help distinguish between two completely different words like in Spanish – *cómo* (how) and *como* (I eat), or French – *cote* (quota), *coté* (respected), *côte* (coast), or *côté* (side). The removal of accents can thus, sometimes lead to the loss of very important information.

It is, therefore, necessary to look at the data and think about the process of its creation. In the past, many SMSs and e-mails were written without diacritics, so some normalization could be useful.

In English, almost no accents are used and their potential removal does not usually cause major problems (e.g., *naïve* → *naive*).

In R, most versions of the `iconv()` function enable diacritics removal by appending `//TRANSLIT` to the output encoding.

```
> x <- "Dobrý den, člověče!"
> iconv(x, from="CP1250", to="ASCII//TRANSLIT")
[1] "Dobry den, clovece!"
```

## 3.12   Normalization

The goal of normalization is to convert tokens that look differently (are, for example, written in a different letter case) but have the same meaning to the same form [263].

### 3.12.1   Case Folding

Case folding is the process of converting all characters of a word to the lower or upper case (the lower case is more common [62]). When two or more words with the same meaning are written with different casing, the number of unique words increases unnecessarily and it is, therefore, favourable to convert them to one form. The reasons why the same word is written with different casing might include the position at the beginning of a sentence, presence in a title, the intention to emphasize a word, etc.

Some words, on the other hand, require capitalization and folding the case in these situations changes the meaning of the word. Typically, names of people, places, companies, etc. are written with the first letter capitalized. For example, *bush* in the middle of a sentence means a small, woody plant while *Bush* would probably mean the former American president (when *Bush* is at the beginning of a sentence, it could be either). After conversion to lower case and not being aware of the context, one cannot distinguish whether a text is about plants or politics according to the occurrence of word *bush*. Abbreviations, although sometimes looking like regular words, are written in the upper case. *SMART* (System for the Mechanical Analysis and Retrieval of Text) when converted to lower case is a synonym for *intelligent*.

Case folding can be problematic in some specific languages. For example, in French, accents are optional for uppercase. The word *PECHE* can be converted to *pêche* which means *fishing*, *peach* or *péché* meaning *sin*.

Capitalization plays an important role in some languages [102]. For example, all nouns are capitalized in German. For instance, *Recht* as a noun means *law* while *recht* as an adjective means *right*. The letter ß also exists only in lower case.

Intelligent case folding is a machine learning problem. As an alternative, a simple approach can convert all words at the beginning of sentences as well as the words from titles and headings to lower case. The case of other words can be retained [6].

## 3.12.2 Stemming and Lemmatization

A *word* is the smallest lexical unit of a language that can be used in isolation. A *morpheme* is the smallest unit of a word that carries some semantic or grammatical meaning. Morphemes typically include prefixes, suffixes, and a root. For example, the word *unexpected* consists of three morphemes – a prefix *un*, a root *expect*, and a suffix *ed*. Some morphemes can be used in a language by themselves (*expect*) while not the others (*un*, *ed*). The former are known as free forms, the latter as bound forms [11].

*Stems* are parts of words that carry the basic meaning. When a stem consists of a single morpheme, it is identical to the root. Free stems can occur alone whereas bound stems cannot.

Words are formed in a word-formation process using several rules. For example, a plural morpheme *s* can be added to a singular noun to create a plural noun (e.g., *book* → *books*). One major morphological operation is the use of *inflection* which is the addition of a plural, possessive, plural+possessive, comparative, superlative, 3rd person singular, past tense, past participle, or present participle forms to a stem. The second type of word-formation operations is *derivation* which includes adding adjective derivational affixes (*day* → *daily*, *depend* → *dependent*, *help* → *helpful* etc.), adverb derivational affixes (*slow* → *slowly*, or *clock* → *clockwise* and many others [126]. Inflection never changes the category of a word (e.g., a noun will be still a noun after adding a suffix) while derivation can change the category (a noun can become an adjective after adding a suffix).

It is obvious that the semantic meaning of some words, when created from the same stem, is very close. When a newspaper article contains one of the words *sport*, *sports*, *sporting*, *sported*, or *sporty*, it is quite likely to be an article from the *sports* category. However, for a computer, all five words are different and to assign an article correctly to the category, all five words need to be connected to it, for example, in a classification rule.

The goal of the procedure known as *stemming* is to convert all different forms of a word to one common form known as the stem.

To find a correct stem, a morphological analysis of the word is often necessary. However, many stemmers (a stemmer is a function or a program performing stemming) employ simple rules to strip affixes. This can lead to a situation that the difference between, for example, a noun and verb disappears – a noun *driver* and verb *driving* will be stemmed to *driv*. We can also see that *driv* is not a correct dictionary word. Some of the stemmers are not constrained to produce only real words [189]. This is usually not considered a deficiency because the word is often used as a feature in a machine learning task and not as a piece of a message for a human. At the same time, when a stem changes with the addition of a suffix (e.g., *index* → *indices*), a simple stemmer will not convert different variants to a common one.

It is obvious that a stemming process is not error-free. The errors include *understemming* (the found stem does not cover all word variants) and *overstemming* (a word is shortened too much so it covers words with different meanings). Understemming is usually achieved by light stemmers that prefer the precision over the recall while aggressive stemmers preferring the recall often overstem the words [39].

The most famous stemming algorithm for English text, belonging to the group of rule-based stemmers, is *Porter's algorithm* [217]. It uses five groups of rules that are successively applied to strings, for example, *SSES* → *SS* (*caresses* → *caress*), *ATIONAL* → *ATE* (*relational* → *relate*), *ALIZE* → *AL* (*formalize* → *formal*, or *AL* → *" "* (*revival* → *reviv*). The rules can contain conditions, like whether the stem contains a vowel, ends with a double consonant, is longer than a given length etc.

Porter [218] also developed a system for defining stemming algorithms. The rules of stemming algorithms are expressed in a natural way in a simple language known as *Snowball*. A Snowball compiler then translates a Snowball definition into an equivalent program in ANSI C or Java.

Dictionary stemmers need a dictionary containing all variants (inflected forms) of a word together with their stems. This method is not suitable for languages with many words and for heavily inflected languages [140].

Rule based and dictionary stemmers require that the rules be constructed by linguistic experts for a specific language. As they are not able to handle situations which are not covered by the rules or the dictionary, statistical or machine learning approaches would be better in some situations. For example, the algorithm of Bacchcin, Ferro, and Melucci et al. [15] estimates the most probable split into the stem and suffix from all possible splits for each word in the full corpus. In [39], the authors clustered together words occurring in similar contexts and sharing a common prefix of at least a minimum length. Such words are expected to be semantically and lexically similar; this can be enough to replace them by one form. They are, however, taken as the training data for a classifier which learns stemming rules so that processing unseen data is possible.

*Lemmatization* converts a word to its dictionary form known as the *lemma*. To be able to perform such a conversion, deeper knowledge than that required for stemming is needed. A lemmatizer usually needs a complete vocabulary and morphological analysis to correctly lemmatize words [185]. The words like *better* or *thoughts* can be converted to their correct forms *good* and *think*, which would not be achieved by a stemmer. Attempts to train a lemmatizer, like in [14, 136], have also been made but they are not error free.

### 3.12.3 Spelling Correction

Mistypings are common problems in any written text. The problem is that they bring noise to the data which can complicate the achievement of satisfactory results in some tasks. *Spelling correction* is a solution to this problem. Generally, the process of spelling correction involves detection of an error, generation of candidate corrections, and ranking of candidate corrections [155].

Mistypings can be detected by comparing each word to an existing dictionary or looking at every character n-gram (n is usually 1, 2, or 3) to determine whether such an n-gram exists for a given language.

When an error is found, our response is to correct it. The correction can be done with or without taking into consideration, the context of the error. The methods which do not take the context into consideration suppose that most of the errors are caused by insertion, deletion, substitution, and transposition; they appear in the middle of a word, can be related to a computer keyboard arrangement, etc.

Sometimes, knowledge of the context in which the error appears can help in choosing the right candidate. The context can also help in finding errors when a regular but incorrect word is used instead of a correct word. For example, in the sentence *Three are two people form Prague here*, only regular English words are used, but *three* and *form* should be replaced with *there* and *from*. The context specific spelling correction algorithms can rely on word association (which candidate correction has a better semantic fit with the words around the misspelling), word repetition (words occurring multiple times in a text can help in finding appropriate corrections), and topical bias (certain correction candidates could be preferred by considering words that are especially relevant to the topic of the text) [90].

Package `hunspell` provides a functionality that can be used when identifying mistyping and correcting the mistakes. Function `hunspell_parse()` tokenizes a text so it can later be checked for errors. As a parameter, the format of input text can be passed (`text`, `latex`, `man`, `html`, or `xml` formats are supported).

```
> t <- c("How are you these days?",
+        "I'm fine, thnx!",
+        "It is <emph>grrreat</emph>, right?")
> words <- hunspell_parse(t, format="html")
> words
[[1]]
[1] "How"   "are"   "you"   "these" "days"

[[2]]
[1] "I"     "m"     "fine" "thnx"

[[3]]
[1] "It"      "is"       "grrreat" "right"
```

The `hunspell_analyze()` function gives information about stems and suffixes of words. To get only words stems, we can use function `hunspell_stem()`.

```
> print(hunspell_analyze(words[[1]]))
[[1]]
[1] " st:how"

[[2]]
[1] " st:are"

[[3]]
[1] " st:you"

[[4]]
[1] " st:these"

[[5]]
[1] " st:day fl:S"
```

To check whether a text contains misspelling, the `hunspell_check()` function can be called. It returns a logical value for each character string representing the presence of an error.

```
> correct <- hunspell_check(words[[3]])
> names(correct) <- words[[3]]
> correct
```

```
    It     is grrreat   right
  TRUE    TRUE  FALSE    TRUE
```

A wrapper function `hunspell()` parses the text, finds incorrect words and returns a list of these words for each line. It effectively combines `hunspell_parse()` with `hunspell_check()` in a single step.

Function `hunspell_suggest()` provides a list of possible replacements for the supplied words (it is typically used to find alternatives for mistyped words).

```
> hunspell_suggest(words[[3]][!correct])
[[1]]
[1] "great"

[[2]]
[1] "right" "girth"
```

Spell checking using `hunspell` is dictionary based. The list of all available dictionaries is returned by the `list_dictionaries()` function.

```
> list_dictionaries()
[1] "en_AU" "en_CA" "en_GB" "en_US"
```

## 3.13 Annotation

The purpose of annotation is to distinguish tokens that look the same but have a different meaning. The context of a token is usually needed to reveal the correct meaning [263]. Annotation is, in fact, an opposite process to normalization.

### 3.13.1 Part of Speech Tagging

Part of speech classes contain words with the same grammatical function [241]. *Open classes* contain words that carry some important meaning and can accept new words. They typically include nouns, verbs or adjectives. They usually contain more words than the closed classes. The words from *closed classes* are relatively constant and new words are rarely added. They usually have a structural function, providing information on the syntactic relations existing among open-class words. Close classes usually contain prepositions, conjunctions, pronouns, or determiners [116]. Part of speech categories used in the famous Penn Treebank project can be found in Table 3.2. The Penn Treebank produced approximately 7 million words of part-of-speech tagged text between the years 1989 and 1996.

**Table 3.2:** An alphabetical list of part of speech tags from the Penn Treebank Project [259].

| Tag | Description |
| --- | --- |
| CC | Coordinating conjunction |
| CD | Cardinal number |
| DT | Determiner |
| EX | Existential there |
| FW | Foreign word |
| IN | Preposition or subordinating conjunction |
| JJ | Adjective |
| JJR | Adjective, comparative |
| JJS | Adjective, superlative |
| LS | List item marker |
| MD | Modal |
| NN | Noun, singular or mass |
| NNS | Noun, plural |
| NNP | Proper noun, singular |
| NNPS | Proper noun, plural |
| PDT | Predeterminer |
| POS | Possessive ending |
| PRP | Personal pronoun |
| PRP$ | Possessive pronoun |
| RB | Adverb |
| RBR | Adverb, comparative |
| RBS | Adverb, superlative |
| RP | Particle |
| SYM | Symbol |
| TO | to |
| UH | Interjection |
| VB | Verb, base form |
| VBD | Verb, past tense |
| VBG | Verb, gerund or present participle |
| VBN | Verb, past participle |
| VBP | Verb, non-3rd person singular present |
| VBZ | Verb, 3rd person singular present |
| WDT | Wh-determiner |
| WP | Wh-pronoun |
| WP$ | Possessive wh-pronoun |
| WRB | Wh-adverb |
| . | Sentence-final punctuation |
| , | Comma |
| : | Colon, semi-colon |
| ( | Left bracket character |
| ) | Right bracket character |
| " | Straight double quote |
| ' | Left open single quote |
| " | Left open double quote |
| ' | Right close single quote |
| " | Right close double quote |

Knowing a part of speech for the words contained in a document might help in extracting information-rich keywords from sentences [146], plagiarism detection [124], named entity recognition [168] and others. Part of speech tagging is also an important part of other tasks like syntactic parsing and machine translation [43].

Successful automatic part of speech taggers are usually built for resource-rich languages in a supervised way [288]. Rule-based taggers use a set of rules used to assign a correct part of speech tag. Statistical taggers try to assign a sequence of tags having the highest probability give by a language model to a text [37].

The package RDRPOSTagger performs part of speech tagging and morphological tagging based on the Ripple Down Rules-based Part-Of-Speech Tagger (RDRPOS). It contains pre-trained models for 45 languages.

The universal annotation (UniversalPOS) uses a reduced set of part of speech tags that is consistent across languages. Another approach (POS) uses an extended language specific tagset. The third option (MORPH) is very detailed morphological annotation. To find out which languages these options are available for, the rdr_available_models() can be used.

In order to annotate a text, a tagger needs to be created using the function rdr_model(). Its parameters are a language and the type of annotation. The function rdr_pos() then takes the created model of a tagger and text, performs part of speech tagging, and returns information about words and assigned tags.

```
> library(RDRPOSTagger)
> text <- "This book is about text mining and
            machine learning."
> tagger <- rdr_model(language="English",
+                     annotation="UniversalPOS")
> rdr_pos(tagger, x=text)
   doc_id token_id    token   pos
1      d1        1     This  PRON
2      d1        2     book  NOUN
3      d1        3       is   AUX
4      d1        4    about   ADP
5      d1        5     text  NOUN
6      d1        6   mining  NOUN
7      d1        7      and CCONJ
8      d1        8  machine  NOUN
9      d1        9 learning  VERB
10     d1       10        . PUNCT
> tagger <- rdr_model(language="English",
+                     annotation="POS")
> rdr_pos(tagger, x=text)
   doc_id token_id    token pos
```

```
1     d1       1     This  DT
2     d1       2     book  NN
3     d1       3       is  VBZ
4     d1       4    about  IN
5     d1       5     text  NN
6     d1       6   mining  NN
7     d1       7      and  CC
8     d1       8  machine  NN
9     d1       9 learning  VBG
10    d1      10        .  .
> text <- "Este libro está escrito en inglés."
> tagger <- rdr_model(language="Spanish",
+                           annotation="MORPH")
> rdr_pos(tagger, x=text)
  doc_id token_id   token      pos
1     d1        1    Este   DDOMSO
2     d1        2   libro   NCMS000
3     d1        3    está   VAIP3SO
4     d1        4 escrito   VMP0OSM
5     d1        5      en   SPS00
6     d1        6  inglés   NCMS000
7     d1        7       .        Fp
```

### 3.13.2  Parsing

The goal of *syntactic parsing* is to find out whether an input sentence is in a given language or to assign the structure to the input text [206]. In order to be able to assign the structure, the grammar of a language is needed. Since it is generally not possible to define rules that would create a parse for any sentence, statistical or machine learning parsers are very important [54].

Complete parsing is a very complicated problem because ambiguities often exist. In many situations, it is enough to identify only unambiguous parts of texts. These parts are known as chunks and they are found using a *chunker* or *shallow parser*. Shallow parsing (chunking) is thus a process of finding non-overlapping groups of words in text that have a clear structure [2].

Parsing in R can be achieved with the udpipe() function from the package udpipe. For more details, see the example at the end of this chapter.

```
> udpipe("There is a man in the small room.",
+        object="english")[, c("token",
+                              "token_id",
```

```
+                                   "head_token_id",
+                                   "dep_rel")]
   token token_id head_token_id dep_rel
1 There         1             2    expl
2    is         2             0    root
3     a         3             4     det
4   man         4             2   nsubj
5    in         5             8    case
6   the         6             8     det
7 small         7             8    amod
8  room         8             4    nmod
9     .         9             2   punct
```

The function returns information about dependency relations between the tokens in the text. Information from the above example can be visualized as in Figure 3.7.

The relation types are from the taxonomy of grammatical relations known as Universal Stanford Dependencies [67] and include the following: clausal modifier of noun (`acl`), adverbial clause modifier (`advcl`), adverbial modifier (`advmod`), adjectival modifier (`amod`), appositional modifier (`appos`), auxiliary (`aux`), case marking (`case`), coordinating conjunction (`cc`), clausal complement (`ccomp`), classifier (`clf`), compound (`compound`), conjunct (`conj`), copula (`cop`), clausal subject (`csubj`), unspecified dependency (`dep`), determiner (`det`), discourse element (`discourse`), dislocated element (`dislocated`), expletive (`expl`), fixed multiword expression (`fixed`), flat multiword expression (`flat`), goes with (`goeswith`), indirect object (`iobj`), list (`list`), marker (`mark`), nominal modifier (`nmod`), nominal subject (`nsubj`), numeric modifier (`nummod`), object (`obj`), oblique nominal (`obl`), orphan (`orphan`), parataxis (`parataxis`), punctuation (`punct`), overridden disfluency (`reparandum`), root (`root`), vocative (`vocative`), and open clausal complement (`xcomp`).

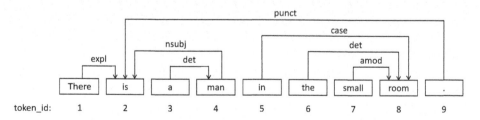

**Figure 3.7:** An example of a parsed sentence.

# 3.14   Calculating the Weights in the Bag-of-Words Model

When it is clear which features will be used to describe the processed document, it is necessary to quantify them. Every feature in each document is then given a weight that somehow represents its importance. The value is generally calculated based on the number of occurrences of words in every document and in the entire collection. The weight $w_{ij}$ of every term $i$ in document $j$ is given by three components [236]:

■  a *local weight* $lw_{ij}$ representing the frequency in every single document,

■  a *global weight* $gw_i$ reflecting the discriminative ability of the term, based on the distribution of the term in the entire document collection, and

■  a *normalization factor* $n_j$ correcting the impact of different document lengths.

The weight $w_{ij}$ is then calculated according to the following formula:

$$w_{ij} = \frac{lw_{ij} * gw_i}{n_j} \qquad (3.4)$$

The local weight is the only mandatory weight which needs to be determined. The other components do not have to be used at all.

## 3.14.1   Local Weights

A local weight quantifies the importance of a term *in a single document*, not considering the other documents. It is therefore based on the frequency of the occurrence of the term in the specific document. The higher the frequency, the higher is the weight.

In some situations, it is only important whether a word is present or absent in a document. The weight is therefore *binary*, also known as *term presence*. Value 0 means that a documents does not contain a term, value 1 says that the word is there, irrespective of its frequency. It is considered to be a baseline to measure the usefulness of a term [252].

The most straightforward method of local term weighting is to simply use the *term frequencies* as weights. If a term is not in a document, the value of the weight is 0, if it is there once, the value is 1, for a double occurrence the value is 2 and so on. Thus, when a term appears ten times in a document it is twice as important as a term appearing five times and ten times more important than a term that is present only once.

Such a direct proportionality, however, does not accurately represent the reality in some situations. When document A contains a certain word five times and the other document B only once, document A is definitely more relevant, for example, in an information retrieval task. But is it really five times more important? Not necessarily. Consider another situation. When retrieving documents using a query consisting of three words ($q = \{w_1, w_2, w_3\}$), a document A containing word $w_1$ four times and words $w_2$ and $w_3$ zero times (the sum of weights of the words shared by the document and the query will be 4) will be more relevant that a document B containing all words once (the sum of weights will be 3). Intuitively, we would expect that document B should be more relevant because it contains more query terms [252].

To decrease the impact of high occurrences of a term, a few weights were developed. Buckley, Allan and Salton [41] and Dumais [77] proposed an application of *logarithm* so the weight grows less than linearly. The *augmented normalized term frequency* puts the values to a certain range, $k$ to 1 (often, $k = 0.5$) [235]. A single occurrence of a given term has some minimal weight and any additional occurrence increases it up to 1. This can be understood as some kind of normalization based on the maximal frequency of a term in a document.

For a detailed description of the mentioned and some other local weights, see Table 3.3. A graph demonstrating how the values of local weights grow, together with increasing term frequencies is contained in Figure 3.8.

## 3.14.2 Global Weights

The goal of global weights is to decrease the importance of terms that are too common. Some English words, like *the*, appear in almost every English text. Thus, this word has no impact on, for example, the categorization of a newspaper article under the *sports* or *politics* category. Neither does it contribute much to determining a similarity between two documents. Its influence on the results of information retrieval are also negligible. We, therefore, want to decrease the weight of such words and give a higher weight to words that do not appear so often so they have a higher *discriminative* power. This means that we need to consider the distribution of words in the entire document collection.

The *Inverse Document Frequency* (IDF) [227] is probably the most popular global weight. It is the logarithm of the inverse probability that a term appears in a random document. When a term is contained in every document, the probability of its appearance is 1 and the logarithm equals to 0. This means that this term has no impact on any subsequent calculation. Inverse document frequency, together with term frequency, is probably the most popular weighting scheme known as *term frequency-inverse document frequency* or simply *tf-idf*.

A probabilistic version of the IDF is the logarithm of the odds that a term appears in a random document [49]. Entropy is the most sophisticated [181] information-theoretic measure [77] taking values between 0 and 1. Some

**Table 3.3:** Calculating the local weight for term $i$ in document $j$ where $f_{ij}$ is the frequency of term $i$ in document $j$, $a_j$ is $average(f_j)$, $x_j$ is $max(f_j)$, $k$ is a user specified constant between 0 and 1, $l_{avg}$ is the average document length, $l_j$ is the length of document $j$.

| Weight name | Calculation |
|---|---|
| Binary (Term Presence) | $lw_{ij} = 0$ if $f_{ij} = 0$ <br> $lw_{ij} = 1$ if $f_{ij} > 0$ |
| Term Frequency (TF) | $lw_{ij} = f_{ij}$ |
| Squared TF | $lw_{ij} = f_{ij}^2$ |
| Thresholded TF | $lw_{ij} = 0$ if $f_{ij} = 0$ <br> $lw_{ij} = 1$ if $f_{ij} = 1$ <br> $lw_{ij} = 2$ if $f_{ij} >= 2$ |
| Logarithm | $lw_{ij} = 0$ if $f_{ij} = 0$ <br> $lw_{ij} = \log\left(f_{ij} + 1\right)$ if $f_{ij} > 0$ |
| Alternate Logarithm | $lw_{ij} = 0$ if $f_{ij} = 0$ <br> $lw_{ij} = 1 + \log f_{ij}$ if $f_{ij} > 0$ |
| Normalized Logarithm | $lw_{ij} = 0$ if $f_{ij} = 0$ <br> $lw_{ij} = \frac{1 + \log f_{ij}}{1 + \log a_j}$ if $f_{ij} > 0$ |
| Augmented Normalized TF | $lw_{ij} = 0$ if $f_{ij} = 0$ <br> $lw_{ij} = k + (1 - k)\left(\frac{f_{ij}}{x_j}\right)$ if $f_{ij} > 0$ |
| Changed-coefficient Average TF | $lw_{ij} = 0$ if $f_{ij} = 0$ <br> $lw_{ij} = k + (1 - k)\frac{f_{ij}}{x_j}$ if $f_{ij} > 0$ |
| Square Root | $lw_{ij} = 0$ if $f_{ij} = 0$ <br> $lw_{ij} = \sqrt{f_{ij} - 0.5} + 1$ if $f_{ij} > 0$ |
| Augmented Logarithm | $lw_{ij} = 0$ if $f_{ij} = 0$ <br> $lw_{ij} = k + (1 - k)\log(f_{ij+1})$ if $f_{ij} > 0$ |
| Augmented Average TF | $lw_{ij} = 0$ if $f_{ij} = 0$ <br> $lw_{ij} = k + (1 - k)\frac{f_{ij}}{a_j}$ if $f_{ij} > 0$ |
| DFR-like Normalization | $lw_{ij} = f_{ij} * \frac{l_{avg}}{l_j}$ |
| Okapi's TF Factor | $lw_{ij} = \frac{f_{ij}}{2 + f_{ij}}$ |

other measures, together with necessary calculation formulae, can be found in Table 3.4.

## 3.14.3 Normalization Factor

The basic idea behind document length normalization is based on two aspects:

■ *Longer documents contain more distinct terms:* This significantly influences the similarity between documents because more matches exist in individual dimensions in the vector space. This is very obvious in an in-

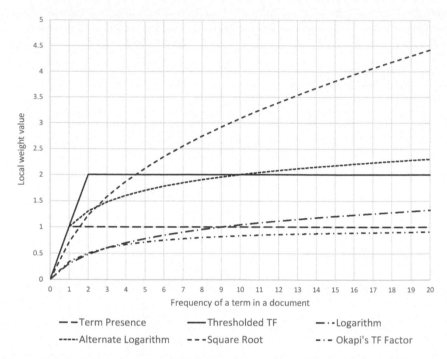

**Figure 3.8:** Dependence between the number of occurrences of a term in a document and the values of selected local weights.

formation retrieval task where there is a higher chance that a query will be matched against longer documents rather than against shorter documents.

■ *Term frequencies in longer documents are higher:* In the case of two documents, one consisting of 50 words where *Prague* appears five times and the other of 1,000,000 words containing *Prague* six times, the first one is definitely related to the Czech Republic more than the second, even though it has a lower frequency of the word *Prague*. When looking at the frequency relatively, the word *Prague* represents one fifth of all words in the first document whereas in the second document, it is a negligible fraction.

In order to calculate the normalization factor, weights of the vector are combined to a single number. These weights are the local weights multiplied by global weights (when used).

The most commonly used normalization technique is *cosine normalization*. It simply transforms a vector to a unit vector (a vector of length 1) by dividing it by its length (calculated as $\sqrt{w_1^2 + w_2^2 + \cdots + w_m^2}$). This simultaneously addresses both reasons for document length normalization – high frequencies as well as

**Table 3.4:** Calculating global weights for term $i$ where $N$ is the number of documents in the collection, $n_i$ is the number of documents containing term $i$ (document frequency), $f_{ij}$ is the frequency of term $i$ in document $j$, $F_i$ is the global frequency of term $i$, and $l_j$ is the length of document $j$.

| Weight name | Calculation |
|---|---|
| None | $gw_i = 1$ |
| Inverse Document Frequency (IDF) | $gw_i = \log \frac{N}{n_i}$ |
| Squared IDF | $gw_i = \log^2 \frac{N}{n_i}$ |
| Probabilistic IDF | $gw_i = \log \frac{N - n_i}{n_i}$ |
| Global frequency IDF | $gw_i = \frac{F_i}{n_i}$ |
| Entropy | $gw_i = 1 + \sum_{j=1}^{N} \frac{\frac{f_{ij}}{F_i} \log \frac{f_{ij}}{F_i}}{\log N}$, |
| Incremented global frequency IDF | $gw_i = \frac{F_i}{n_i} + 1$ |
| Log-global frequency IDF | $gw_i = \log \left( \frac{F_i}{n_i} + 1 \right)$ |
| Square root global frequency IDF | $gw_i = \sqrt{\frac{F_i}{n_i} - 0.9}$ |
| Inverse total term frequency | $gw_i = \log \frac{\sum_{j=1}^{N} l_j}{F_i}$ |

**Table 3.5:** Normalization of weights in document $j$, where $gw_i$ is the value of the global weight of term $i$, $lw_{ij}$ is the value of the global weight of term $i$ in document $j$, $m$ is the number of terms in document $j$, and $f_{ij}$ is the frequency of term $i$ in document $j$.

| Weight name | Calculation |
|---|---|
| None | $n_j = 1$ |
| Cosine | $n_j = \sqrt{\sum_{i=1}^{m} (gw_i * lw_{ij})^2}$ |
| Sum of weights | $n_j = \sum_{i=1}^{m} gw_i * lw_{ij}$ |
| Max weight | $n_j = \max gw_i * lw_{ij}$ |
| Max TF | $n_j = 0.5 + 0.5 * \frac{gw_i * lw_{ij}}{\max gw_i * lw_{ij}}$ |
| Square root | $f_{ij} = \sqrt{gw_i * lw_{ij}}$ |
| Logarithm | $f_{ij} = \log gw_i * lw_{ij}$ |
| Fourth normalization | $n_j = \sum_{i=1}^{m} (gw_i * lw_{ij})^4$ |

a larger number of terms (non-zero elements) in a document; as a result, the normalization factor has a higher value [252]. This kind of normalization also has another positive effect. In order to calculate the cosine similarity of two vectors (see section 12.3.1) only the dot product of these vectors needs to be calculated [185].

Other methods include dividing every vector component by the maximal value in the vector which shrinks the components' values to the range 0 to 1 [216] or scaling to the range 0.5 to 1 by adding one half of the ratio between a vector component and the maximal value to 0.5 [141].

# 3.15 Common Formats for Storing Structured Data

After some of the preprocessing steps mentioned above have been carried out and the values of the document-term matrix components have been calculated, the data is ready for further processing by machine learning algorithms. The data is now in an internal form in the operational memory of the given software package performing the conversion to a structured representation. When the software has the necessary capabilities, data mining algorithms can be applied to the data. When a different data mining software is used (for example, when it implements a specific algorithm with much higher speed of computation), the data usually needs to be stored to a file with the specific structure required by that software and loaded there. This section describes a few common formats for storing structured data which are often used in machine learning.

## 3.15.1 Attribute-Relation File Format (ARFF)

The *Attribute-Relation File Format* (ARFF) is a format developed by the Machine Learning Project at the Department of Computer Science of The University of Waikato for use with their Weka machine learning software [280]. A file having this format consists of a header and data sections.

The header contains the name of a relation and a list of attributes, together with their data types. A relation name as well as an attribute name is a string (double quoted when containing spaces). The data type can be one of the four following types:

- numeric – integer or real numbers

- <nominal-specification> – a list of possible values is enumerated between a pair of curly brackets, comma separated (values containing spaces need to be quoted)

- string – text values

- date [<date-format>] – date and time values, optionally with a specification of how to parse them (*yyyy-MM-dd'T'HH:mm:ss* by default)

The data section contains one instance on every row. An instance is represented by a comma separated list of values in the order given in the header section. Strings and nominal values containing spaces must be quoted; dates are always quoted. Missing values are represented by a single question mark.

An example of an ARFF file containing e-mails to be classified as spam and ham can be found in Figure 3.9. In this example, the e-mail texts are in their original textual form. When the raw text data is transformed to a structured representation, more attributes will appear. This is visible in Figure 3.10.

From Figure 3.10 is obvious that, even in the case of this simple example, most of the values in the vectors are zeros (they are sparse). When a data set contains a number of texts, the vectors are usually very large and the problem is intensified. This also means that the ARFF files tend to occupy a large space on the disks because they store many unnecessary zero values. To overcome this problem, it is possible to store only non-zero values. It is, however, necessary to add an index to each such value. The indices start from 0 and the values are stored in space separated pairs <index> <value>. Every instance is enclosed in a pair of curly brackets. An example of a sparse ARFF file is in Figure 3.11.

Both ARFF and sparse ARFF files can be stored in a XML-based format known as eXtensible attribute-Relation File Format (XRFF). The format is able to include some additional properties of a data set like class attribute specification, attribute weights, and instance weights to its description. In a sparse XRFF file, an index for each value needs to be specified as an attribute of the <value> element. Unlike in the sparse ARFF format, the indices begin with 1. Examples of XRFF files for the e-mails dataset can be found in Figures 3.12 and 3.13.

## 3.15.2 Comma-Separated Values (CSV)

The *comma-separated values* format is a very simple format for structured data interchange.

```
@relation e-mails

@attribute text string
@attribute class {spam, ham}

@data
'Want buy cheap V*I*A*G*R*A?',spam
'The meeting will take place on Monday.',ham
'A new version of our product will be launched next week.',ham
'Need help with transferring funds from Africa',spam
```

**Figure 3.9:** An example of an ARFF file containing e-mails to be classified as either *spam* or *ham*.

```
@relation "e-mails structured"

@attribute Africa numeric
@attribute Need numeric
@attribute V*I*A*G*R*A numeric
@attribute Want numeric
@attribute buy numeric
@attribute cheap numeric
@attribute from numeric
@attribute funds numeric
@attribute help numeric
@attribute transferring numeric
@attribute with numeric
@attribute A numeric
@attribute Monday numeric
@attribute The numeric
@attribute be numeric
@attribute launched numeric
@attribute meeting numeric
@attribute new numeric
@attribute next numeric
@attribute of numeric
@attribute on numeric
@attribute our numeric
@attribute place numeric
@attribute product numeric
@attribute take numeric
@attribute version numeric
@attribute week numeric
@attribute will numeric
@attribute class {spam,ham}

@data
0,0,1,1,1,1,0,0,0,0,0,0,0,0,0,0,0,0,0,0,0,0,0,0,0,0,0,0,spam
0,0,0,0,0,0,0,0,0,0,0,0,0,1,1,0,0,1,0,0,0,1,0,1,0,1,0,0,1,ham
0,0,0,0,0,0,0,0,0,0,0,0,1,0,0,1,1,0,1,1,1,0,1,0,1,0,1,1,1,ham
1,1,0,0,0,0,1,1,1,1,1,0,0,0,0,0,0,0,0,0,0,0,0,0,0,0,0,0,spam
```

**Figure 3.10:** An example of an ARFF file containing e-mails to be classified as *spam* or *ham*. The data is already transformed to a structured representation (bag-of-words model) when no special preprocessing techniques have been applied.

Each line in a CSV file contains one instance. The attributes of each instance are separated by commas and they are always in the same order. When a comma (which normally separates values) or a line break (which generally denotes the end of a record) is contained in a data field, the value should be enclosed between

```
@relation "e-mails sparse"

@attribute Africa numeric
@attribute Need numeric
...
@attribute class {spam,ham}

@data
{2 1,3 1,4 1,5 1,28 spam}
{12 1,13 1,16 1,20 1,22 1,24 1,27 1,28 ham}
{11 1,14 1,15 1,17 1,18 1,19 1,21 1,23 1,25 1,26 1,27 1,28 ham}
{0 1,1 1,6 1,7 1,8 1,9 1,10 1,28 spam}
```

**Figure 3.11:** An example of an ARFF file containing e-mails to be classified as *spam* or *ham*. The data is already transformed to a structured representation (bag-of-words model) when no special preprocessing techniques have been applied. Instead of the sparse vectors that contain many zeros, dense vectors are used.

a pair of double quotes. In such a situation, every double-quote appearing in a field must be escaped by preceding it with another double quote.

Optionally, a CSV file can include a header containing attribute names on the first line. The format of a header is the same as for the rest of the file.

Although the format is standardized [248], many variations based on the above mentioned principles exist. For example, a tabulator can be used to separate the values instead of commas, or a line end different from CRLF can be considered.

An example of a CSV file containing structured representation of e-mails is in Figure 3.14.

### 3.15.3   C5 format

C5 (a UNIX version) and See5 (a Windows version) are very efficient data mining tools, enabling the construction of decision trees or rules and using them for classification and prediction [220].

The data to be processed by the C5 package is divided into several files. They share the same name called a filestem. An extension then defines what aspect of the data is stored in the file named *<filestem>.<extension>*.

A file with an extension *names* defines the attributes and a file with an extension *data* contains the values of the instances to be processed. These two files are required. The other three optional file types contain a data set for testing (extension *test*), instances to be classified (extension *cases*), and information about misclassification costs (extension *costs*).

```
<dataset name="e-mails structured">
  <header>
    <attributes>
      <attribute name="Africa" type="numeric"/>
      <attribute name="Need" type="numeric"/>
      <attribute name="V*I*A*G*R*A" type="numeric"/>
      ...
      <attribute name="class" type="nominal">
        <labels>
          <label>spam</label>
          <label>ham</label>
        </labels>
      </attribute>
    </attributes>
  </header>
  <body>
    <instances>
      <instance>
        <value>0</value>
        <value>0</value>
        <value>1</value>
        ...
        <value>spam</value>
      </instance>
      <instance>
        <value>0</value>
        <value>0</value>
        <value>0</value>
        ...
        <value>ham</value>
      </instance>
      ...
    </instances>
  </body>
</dataset>
```

**Figure 3.12:** An example of an XRFF file containing e-mails to be classified as *spam* or *ham*. The data is already transformed to a structured representation (bag-of-words model) when no special preprocessing techniques have been applied. Instead of the sparse vectors that contain many zeros, dense vectors are used.

The program generates files with extensions *tree* or *rules* containing the generated decision tree or rules.

In the *names* file, specification of a class is the first information. A class can be defined in three different ways:

■ a comma separated list of class labels, e.g., spam, ham.

```
<dataset name="e-mails structured">
  <header>
  ...
  </header>
  <body>
    <instances>
    <instance type="sparse">
      <value index="3">1</value>

      ...

      <value index="5">spam</value>
    </instance>
    <instance type="sparse">
      <value index="1">1</value>
      <value index="5">ham</value>
    </instance>
    </instances>
  </body>
</dataset>
```

**Figure 3.13:** An example of a XRFF file containing e-mails to be classified as *spam* or *ham*. The data is already transformed to a structured representation (bag-of-words model) when no special preprocessing techniques have been applied. Instead of the sparse vectors that contain many zeros, dense vectors are used. The header section is the same as in case of the dense XRFF example.

```
Africa,Need,V*I*A*G*R*A,...,week,will,class
0,0,1,1,1,1,0,0,0,0,0,0,0,0,0,0,0,0,0,0,0,0,0,0,0,0,0,0,spam
0,0,0,0,0,0,0,0,0,0,0,0,0,1,1,0,0,1,0,0,0,1,0,1,0,1,0,0,1,ham
0,0,0,0,0,0,0,0,0,0,0,1,0,0,1,1,0,1,1,1,0,1,0,1,0,1,1,1,ham
1,1,0,0,0,0,1,1,1,1,1,0,0,0,0,0,0,0,0,0,0,0,0,0,0,0,0,0,spam
```

**Figure 3.14:** An example of a CSV file containing e-mails to be classified as *spam* or *ham*. The data is already transformed to a structured representation (bag-of-words model) when no special preprocessing techniques have been applied.

■ the name of a discrete attribute to be used as a class

■ the name of a continuous attribute together with thresholds separating all possible values to a countable number of intervals, e.g., age: 21, 50. defines three classes ($age \leq 21$, $21 < age \leq 50$, and $50 \leq age$)

A list of all attribute names, together with their data types, follows. A data type is separated by a colon from an attribute name. A few data types are available: continuous for numeric values, date for dates (in the form YYYY/MM/DD or YYYY-MM-DD), time for times (in the form HH:MM:SS with values between 00:00:00 and 23:59:59), timestamp combining date and time (in the form YYYY/MM/DD HH:MM:SS or YYYY-MM-DD HH:MM:SS), a comma-separated list of names enumerating the allowable values (e.g., `risk: low, medium, high`), `discrete N` for discrete, unordered values that are assembled from the data itself, `ignore` for attributes to be ignored, and `label` for naming and referring the instances (labels are ignored during processing). Attributes can be also defined by formulae, e.g., `Difference := A - B`.

At the end, optional information that explicitly specifies what attributes should be processed can be included. This either lists the processed attributes as `attributes included: a1, a3, a5.` or contains the attributes that will not be used, like `attributes excluded: a2, a4.`

A *data* file contains the values of training instances. Each instance contains a comma separated list of values for all explicitly-defined attributes. If the classes are listed in the first line of the *names* file, the last value on each line is a class label. A question mark is used to denote a missing or unknown value, N/A denotes a value that is not applicable. The values of the attributes that are calculated from other attributes are naturally not included.

The *test* and *cases* files have the same format as the *data* file. In the *cases* file, the class vales can be unknown.

An example of *names* and *data* file can be found in Figure 3.15.

file emails.names:

```
spam, ham.
Africa  : continuous.
Need    : continuous.
VIAGRA  : continuous.
...
week    : continuous.
will    : continuous.
```

file emails.data:

```
0,0,1,1,1,1,0,0,0,0,0,0,0,0,0,0,0,0,0,0,0,0,0,0,0,0,0,0,spam
0,0,0,0,0,0,0,0,0,0,0,0,1,1,0,0,1,0,0,0,1,0,1,0,1,0,0,1,ham
0,0,0,0,0,0,0,0,0,0,0,1,0,0,1,1,0,1,1,1,0,1,0,1,0,1,1,1,ham
1,1,0,0,0,0,1,1,1,1,1,0,0,0,0,0,0,0,0,0,0,0,0,0,0,0,0,0,spam
```

**Figure 3.15:** An example of files suitable for the C5 package. They describe e-mails to be classified as *spam* or *ham*. The data is already transformed to a structured representation (bag-of-words model) when no special preprocessing techniques have been applied.

## 3.15.4 Matrix Files for CLUTO

CLUTO is a software package for clustering low- and high-dimensional datasets and for analyzing the characteristics of the various clusters. It contains many clustering algorithms, similarity/distance functions, clustering criterion functions, agglomerative merging schemes, methods for effectively summarizing the clusters, and visualization capabilities [141].

A matrix file (*.mat*) contains a dense matrix, sparse matrix, or dense/sparse similarity graph representing the data to be clustered.

The first row of a dense matrix contains the number of rows and columns of the matrix. The remaining rows contain one instance on each row. The rows have a form of space separated floating point values, including non-zero values. A sparse matrix contains the number of rows, columns, and non-zero values of the matrix on the first row. The remaining rows contain one instance on each row. The rows have a form of space separated pairs of indexes and corresponding non-zero values (which do not need to be ordered). The indexes begin with 1.

A dense graph format contains the number of vertices of the graph on the first row. The remaining rows contain an adjacency matrix, which is a matrix $n \times n$ where $n$ is the number of vertices. An element on position $i$ and $j$ is a floating point value representing the similarity of the $i$th and the $j$th vertex of the graph. A sparse graph format contains the numbers of vertices and edges of the graph on the first row. The remaining rows contain an adjacency matrix in the form of index-value pairs (the indexes begin with 1).

Additionally, CLUTO accepts an *rlabel* file containig labels for the rows of the data matrix, *clabel* file with labels for the columns of the data matrix, and *rclass* file containing class labels for the rows of the data matrix for an external cluster evaluation.

## 3.15.5 SVMlight Format

The format was developed by Thorsten Joachims to describe the data to be processed by his famous SMV$^{light}$ package that implements the support vector machine (SVM) algorithm [135].

Every line in a file represents one training example. The first value is a target and can be one of the following values: $+1$, $-1$, $0$, or a floating point number. The classes are normally labeled as $-1$ and $+1$. A class label of 0 indicates that this example should be classified using transduction. The target can also contain a real number which is used in the regression mode.

A list of feature-value pairs follows. A feature is described by its number, where the first feature has number 1. A value is a floating point number. The target value and each of the feature-value pairs, which must be ordered by increasing feature numbers, are separated by a space character. Features with zero values do not have to be included, which is common for sparse formats file *emails.mat* (sparse format):

```
4 28 29
3 1 4 1 5 1 6 1
13 1 14 1 17 1 21 1 23 1 25 1 28 1
12 1 15 1 16 1 18 1 19 1 20 1 22 1 24 1 26 1 27 1 28 1
1 1 2 1 7 1 8 1 9 1 10 1 11 1
```

file *emails.rlabelfile*:

```
spam
ham
ham
spam
```

file *emails.clabel*:

```
Africa
Need
VIAGRA
...
week
will
```

**Figure 3.16:** An example of files suitable for the CLUTO package. They describe e-mails to be classified as *spam* or *ham*. The data is already transformed to a structured representation (bag-of-words model) when no special preprocessing techniques have been applied.

Every row can be followed by a string starting with #. It is then used as additional information to a user defined kernel.

An example of a SVMlight file can be found in Figure 3.17.

```
1 3:1 4:1 5:1 6:1
-1 13:1 14:1 17:1 21:1 23:1 25:1 28:1
-1 12:1 15:1 16:1 18:1 19:1 20:1 22:1 24:1 26:1 27:1 28:1
1 1:1 2:1 7:1 8:1 9:1 10:1 11:1
```

**Figure 3.17:** An example of a file in the SVMlight format describing e-mails to be classified as *spam* or *ham*. The data is already transformed to a structured representation (bag-of-words model) when no special preprocessing techniques have been applied.

## 3.15.6   Reading Data in R

There are a few packages which provide functions implementing the reading of data from common formats.

The `read.arff()` function from package `RWeka` reads data from Attribute-Relation File Format (ARFF) files. The only input parameter is a character string with the name of the file or a connection. The function returns a data frame containing the data.

To read data in a tabular format, including CSV files, the function `read.table()` can be used. The function reads a file in table format and creates a data frame from it. The lines correspond to individual instances and fields on each line correspond to attributes of the instances. Using input parameters, a user can specify whether the file contains a header, identify the field separator character, specify how strings are quoted, which character is used for decimal points, what is the maximum number of rows to read or how many lines of the data to skip. The functions `read.csv()` and `read.csv2()`, intended for reading the content of CSV files, are identical to `read.table()` but have different default values for their parameters.

Using the function `C45Loader()` from package `RWeka`, data in the C4.5 format (which is identical to the C5 format described above) can be read.

Package `skmeans` provides function `readCM()` that reads a sparse matrix in the CLUTO format and returns a triplet matrix (values are represented by triplets – row index, column index, value) defined by the indexR packages!slamslam package.

To save data in a format of R, the function `save()` can be used. It accepts a list of whatever objects are written to a file (with an extension *.RData*) with the name given as the parameter, `file`. These objects can be read later and the structures in the memory recreated using function `load()` where the parameter is the file name with stored objects.

## 3.16   A Complex Example

Like in almost every major statistical computing product, R also offers text mining capabilities. A well-known text mining framework, `tm` is an open source package providing the basic infrastructure necessary to organize, transform, and analyze textual data [86].

A document collection can be represented by an object of class `VCorpus` (Volatile Corpus, a default implementation, documents are stored in the volatile memory) or `PCorpus` (Permanent Corpus, where the documents are stored outside an R program). The first parameter of the methods creating a corpus (`VCorpus()` or `PCorpus()`) is a `Source` class object which specifies the source of the data.

A number of methods, each handling data in different formats, may be used to create the corpus. The complete list of available sources is returned by the `getSources()` function. It includes, for example:

■ `VectorSource()` – each element of a vector is interpreted as a document.

- `DataFrameSource()` – the corpus is created for a data frame where the first column is names `doc_id` (a unique string identifier) and the second column is `text` (a UTF-8 encoded string).

- `DirSource()` – each file in a directory is treated as a document.

- `URISource()` – the data from a specified URI is used to form a document.

The data from a specified source is read by a function known as a reader. The behaviour of the reader is influenced by the parameter, `readerControl`. This is a list with components `reader`, specifying a particular reader to be used, and `language`, defining the language of the documents. The readers (their complete list is returned by the function `getReaders()`) include, for example, `readDataframe` (reads data from a data frame), `readDOC` (reads data from a Microsoft Word document), `readPDF` (reads data from a Portable Document Format document), `readPlain` (reads texts from a list with a `content` component), or `readXML` (reads an XML document). In addition to an object containing the data, all functions also accept additional information suitable for a specific data source (for example, the name of a PDF extraction engine for reading PDF documents). A language (en by default) is passed as an IETF language tag.

The documents in a corpus can be modified using the `tm_map()` function. This function applies another function known as a transformation to all documents in a corpus. The list of transformations, which can be retrieved by the `getTransformations()` functions, includes:

- `removeNumbers()` – removes numbers; a logical parameter, ucp, specifies whether to use Unicode character properties for determining digit characters.

- `removePunctuation()` – removes punctuation. With the parameter, `preserve_intra_word_contractions`, a user specifies whether intra-word contractions should be kept; the `preserve_intra_word_dashes` parameter controls the removal of intra-word dashes. The parameter ucp determines the punctuation characters. They include ! " # $ % & ' ( ) * + , - . / : ; < = > ? @ \ [ ] ^ _ ` { | } ~ when FALSE (by default), or the Unicode punctuation characters when TRUE.

- `removeWords()` – returns a document without the words specified as the second parameter (or a parameter named `words`). This function is often used to remove stop words. The list can be returned by the `stopwords()` function with a parameter specifying the language.

- `stemDocument()` – performs stemming using using the Porter's stemming algorithm. An optional parameter, `language`, can be used to choose a proper Snowball stemmer.

■ `stripWhitespace()` – replaces multiple whitespaces with a single space in a document.

Custom transformations can be created with the `content_transformer()` function. This function accepts as a parameter, another function which modifies the content of an object with implemented content getter (`content`) and setter (`content<-`) functions. An existing function as well as a function defined by a programmer can be used, as demonstrated below:

```
# an existing function
c <- tm_map(c, content_transformer(tolower))

# a defined function replacing all e-mail addresses
# with "--email--"
email.replace <- content_transformer(
        function (x) gsub("([\\w.]+)@([\\w.]+\\.[a-z]+)",
                          "--email--",
                          x,
                          perl=TRUE)
        )
c <- tm_map(c, content_transformer(email.replace))
```

The function, `tm_filter()`, enables the selection of only the documents which meet certain criteria. It returns a corpus containing documents where a function passed as the second parameter returns TRUE. The function `tm_index()` works on the same principle but returns only a logical value for each document.

To save a corpus on a disk, the function `writeCorpus()` can be used. It takes a corpus as an input and writes the content (applies `as.character()` to each document) to files in a given location. Filenames are generated from document IDs or from character strings passed as an optional parameter.

To create a document-term matrix (documents are in rows) or term-document matrix (terms are in rows), the functions `DocumentTermMatrix()` and `TermDocumentMatrix()` can be used. Both functions implement sparse matrices. The `as.matrix()` function can be used to display a full matrix. A sample is displayed by the function `inspect()`.

To control the process of calculating the values (weights of the terms) in the matrices, a named list, `control`, with options is supplied. Local options are evaluated for each document and global options are evaluated once for the entire matrix.

The local options include:

■ `tokenize` – a function tokenizing a document into single tokens. The function can be one of predefined functions or a user defined one.

- `tolower` – a logical value indicating whether characters should be translated to lower case or a custom function converting characters to lower case.

- `language` – the language (preferably as a IETF language tag) to be used for the removal of stop words and stemming.

- `removePunctuation` – a logical value indicating whether punctuation characters should be removed, a function performing punctuation removal, or a list of arguments for `removePunctuation()`.

- `removeNumbers` – a logical value indicating whether numbers should be removed or a function implementing number removal.

- `stopwords` – a Boolean value indicating stop word removal using default language specific stop words lists, a character vector holding custom stop words, or a custom function for stop words removal.

- `stemming` – a Boolean value indicating whether tokens should be stemmed or a custom stemming function.

- `dictionary` – a character vector containing acceptable terms; other terms in the documents will be ignored.

- `bounds` – a list with a tag `local` containing an integer vector of length two. The first value specifies the minimal number of occurrences of a term in a document, the second value defines the maximal number of occurrences of the term in a document.

- `wordLengths` – an integer vector of length two. The value defines the minimal and maximal lengths of a term.

Global options include:

- `bound` – a list with tag `global` which contains an integer vector with two elements. The elements are the lower and upper bounds for the document frequency of a term. Terms that are out of the bounds are eliminated.

```
# eliminating terms that appear less than twice
tdm <- TermDocumentMatrix(
        c,
        control=list(bounds=list(global=c(2, Inf)))
        )
```

∎ weighting – a weighting function. The tm package provides weightTf (term frequency, a default option), weightTfIdf (term frequency – inverse document frequency), weightBin (binary weighting, term presence/absence), and weightSMART options. The weightSMART option defines weights in SMART notation. Three characters in a string passed as a parameter specify the local and global components and a normalization factor. The local weights include term frequency (n), logarithm (l), augmented normalized term frequency (a), term presence (b), and normalized logarithm also known as log average (L). The global options include no weighting (n), inverse document frequency (t), and probabilistic IDF (p). The vectors can be normalized using the cosine normalization (c), pivoted normalization (u), byte size (b), or do not have to be normalized (n).

A useful function which reduces the sparsity of a matrix representing the documents is removeSparseTerms(). It removes terms that have a higher percentage of empty (zero) values in a document term matrix than the parameter (a value between 0 and 1).

There are a few functions which provide some basic information about the content of a term-document matrix or document-term matrix, which is a parameter of these functions. These functions include Docs() (returns document IDs), Terms() (returns a list of terms), nDocs() (returns the number of documents), and nTerms() (returns the number of documents).

The function findMostFreqTerms() calculates and returns the frequencies of words in every document or group of documents (given by the INDEX object) ordering the frequency in descending order. A maximum number of words (parameter n) can be given. The documents are in the form of a term-document-matrix, document-term-matrix, or a vector of frequencies returned by termFreq() (which calculates term frequencies for a single document).

The function plot() visualizes correlations between the terms of a term-document matrix (it requires the Rgraphviz library) in the form of a graph where the vertices (terms) are connected by edges (associations). Terms to be plotted (20 terms, randomly chosen by default) and correlation threshold are some of the possible parameters.

To visualize Zipf's law and Heaps' law based on the values contained in a document-term matrix or term-document matrix with unweighted term frequencies, tm provides the functions Zipf_plot() and Heaps_plot(). A user can specify the type of plot to be drawn and further graphical parameters to be used for plotting.

An example of using the tm package for analyzing a corpus and for preparing a document term matrix follows.

```
> d <- c("The Vegas Golden Knights ice hockey team
+          achieved the milestone of most home wins.",
+        "Germany won 1-0 over Argentina in the FIFA Wold
+         Cup finals.",
+        "Someone has won tonight's USD 10 million jackpot
+         in the National Lottery.",
+        "A man has won a USD 10 million jackpot on
+         a Las Vegas slot machine.")

> # creating a corpus from a character vector
> c <- VCorpus(VectorSource(d))

> # looking at some basic information about the corpus
> c
<<VCorpus>>
Metadata:  corpus specific: 0, document level (indexed): 0
Content:  documents: 4

> # looking at the text of the second document
> c[[2]]$content   # same as as.character(c[[2]])
[1] "Germany won 1-0 over Argentina in the FIFA Wold
Cup finals."

> # finding only documents that contain the word "Vegas"
> c2 <- tm_filter(c,
+              FUN=function(x)
+                  any(grep("Vegas", content(x))))
> lapply(c2, as.character)
$'1'
[1] "The Vegas Golden Knights ice hockey team achieved the
milestone of most home wins."

$'4'
[1] "A man has won a USD 10 million jackpot on a Las
Vegas slot machine."

# finding indices of documents that contain the word "Vegas"
> tm_index(c, FUN=function(x)
+                  any(grep("Vegas", content(x))))
[1]  TRUE FALSE FALSE  TRUE

> # writing the documents from the filtered corpus
> # to a directory "filtered_corpus"
```

```
> writeCorpus(c2, path="filtered_corpus")

> # removing numbers from the documents
> c <- tm_map(c, removeNumbers)

> # looking at the text of all documents
> lapply(c, as.character)
$'1'
[1] "The Vegas Golden Knights ice hockey team achieved
the milestone of most home wins."

$'2'
[1] "Germany won - over Argentina in the FIFA Wold Cup
finals."

$'3'
[1] "Someone has won tonights USD  million jackpot in
the National Lottery."

$'4'
[1] "A man has won a USD  million jackpot on a Las
Vegas slot machine."

> # removing punctuation
> c <- tm_map(c, function(x)
+               removePunctuation(
+                 x,
+                 preserve_intra_word_contractions=FALSE))

> # performing stemming
> c <- tm_map(c, stemDocument)

> # looking at the text of all documents after
> # transformations
> lapply(c, as.character)
$'1'
[1] "The Vega Golden Knight ice hockey team achiev the
      mileston of most home win"

$'2'
[1] "Germani won over Argentina in the FIFA Wold Cup final"

$'3'
```

```
[1] "Someon has won tonight USD million jackpot in the
    Nation Lotteri"

$'4'
[1] "A man has won a USD million jackpot on a Las Vega
    slot machin"

> # generating a document-term matrix
> dtm <- DocumentTermMatrix(c)

# display detailed information on the document-term matrix
inspect(dtm)

<<DocumentTermMatrix (documents: 4, terms: 32)>>
Non-/sparse entries: 41/87
Sparsity            : 68%
Maximal term length: 9
Weighting           : term frequency (tf)
Sample              :
      Terms
Docs achiev argentina cup has jackpot million the usd
   1      1         0   0   0       0       0   2   0
   2      0         1   1   0       0       0   1   0
   3      0         0   0   1       1       1   1   1
   4      0         0   0   1       1       1   0   1
      Terms
Docs vega won
   1    1   0
   2    0   1
   3    0   1
   4    1   1

> # displaying the document-term matrix as a matrix
> as.matrix(dtm)

Docs achiev argentina cup fifa final ...
   1      1         0   0    0     0 ...
   2      0         1   1    1     1 ...
   3      0         0   0    0     0 ...
   4      0         0   0    0     0 ...

> # looking at some properties of the document-term matrix
> Docs(dtm)
```

```
[1] "1" "2" "3" "4"
> nDocs(dtm)
[1] 4
> Terms(dtm)
 [1] "achiev"    "argentina" "cup"       "fifa"
 [5] "final"     "germani"   "golden"    "has"
 [9] "hockey"    "home"      "ice"       "jackpot"
[13] "knight"    "las"       "lotteri"   "machin"
[17] "man"       "mileston"  "million"   "most"
[21] "nation"    "over"      "slot"      "someon"
[25] "team"      "the"       "tonight"   "usd"
[29] "vega"      "win"       "wold"      "won"
> nTerms(dtm)
[1] 32


> # looking at 5 most frequent terms in the entire collection
> findMostFreqTerms(dtm, INDEX=rep(1, times=nDocs(dtm)),n=5)
$`1`
     the     won     has jackpot million
       4       3       2       2       2


> # removing columns that have more than a half of
> # zero elements
> dtm2 <- removeSparseTerms(dtm, 0.51)
> as.matrix(dtm2)
     Terms
Docs has jackpot million the usd vega won
   1   0      0       0   2   0    1   0
   2   0      0       0   1   0    0   1
   3   1      1       1   1   1    0   1
   4   1      1       1   0   1    1   1


> # storing data to a CSV file to be used later
> m <- as.matrix(dtm2)
> write.csv(m, file="data.csv")
```

As an alternative to the package tm, a natural language processing toolkit udpipe can be used. The package provides tokenization, part of speech tagging, lemmatization, and dependency parsing capabilities. It also enables training annotation models, converting texts to a document-term matrix, and some basic operations with the matrix.

The function udpipe() analyzes text and returns a data frame where one row corresponds to one token in the data. The fields contain a document identifier (doc_id), a paragraph identifier unique within each document (paragraph_id), a sentence identifier unique within each document (sentence_id), the text of the sentence (sentence), integer indices indicating the beginning and end of the token in the original text (start, end), a row identifier unique within the document (term_id), a token index starting at 1 for each sentence (token_id), the token (token), the lemma of the token (lemma), the universal and treebank-specific part of speech tags of the token (upos, xpos), the morphological features of the token separated by | (feats), indicating what token_id in the sentence the token is related to (head_token_id) and what type of relation it is (dep_rel), enhanced dependency graph (deps), and information about spaces in a token used to reconstruct the original text (misc).

The function document_term_frequencies(d) calculates the number of times a term occurs per document. The input is a data frame, data table, or character vector (it needs to be split first according to the submitted optional parameter split). The output is a data table with columns doc_id, term, and freq indicating how many times a term occurred in each document.

A data table returned by the function document_term_frequencies() can be passed to document_term_frequencies_statistics(). This function adds term frequency normalized by the number of terms in a document, inverse document frequency, and Okapi BM25 statistics to the output.

The output from the document_term_frequencies(d) function can be converted to a document-term matrix using the function document_term_matrix(). The matrix is a sparse object of class dgCMatrix with the documents in the rows and the terms in the columns. Optionally, a vocabulary can be supplied using the vocabulary parameter. In such a case, only terms which are in the vocabulary will be in the matrix.

Two sparse matrices can be joined using dtm_cbind() or dtm_rbind() (by columns or by rows).

The functions dtm_colsums() and dtm_rowsums(dtm) return column sums and row sums for a document term matrix.

Using a function dtm_remove_lowfreq(), terms occurring with low frequency and documents with no terms are removed from a document-term matrix. The minimal frequency is given by the minfreq parameter. An optional parameter, maxterms, indicates the maximum number of terms to be kept in the matrix. Similarly, the function dtm_remove_tfidf() removes terms with low tf-idf frequencies.

To eliminate specific terms, the function dtm_remove_terms() can be used. The terms are passed as the terms parameter.

```
> library(udpipe)
> udpipe("Machine learning is great. Learn it!",
+         object="english")
  doc_id paragraph_id sentence_id                    sentence
```

```
1   doc1            1       1 Machine learning is great.
2   doc1            1       1 Machine learning is great.
3   doc1            1       1 Machine learning is great.
4   doc1            1       1 Machine learning is great.
5   doc1            1       1 Machine learning is great.
6   doc1            1       2           Learn it!
7   doc1            1       2           Learn it!
8   doc1            1       2           Learn it!
  start end term_id token_id   token    lemma   upos xpos
1     1   7       1        1 Machine  machine   NOUN   NN
2     9  16       2        2 learning learning  NOUN   NN
3    18  19       3        3      is       be    AUX  VBZ
4    21  25       4        4   great    great    ADJ   JJ
5    26  26       5        5       .        . PUNCT    .
6    28  32       6        1   Learn    Learn   VERB   VB
7    34  35       7        2      it       it   PRON  PRP
8    36  36       8        3       !        ! PUNCT    .
                                                    feats
1                                            Number=Sing
2                                            Number=Sing
3 Mood=Ind|Number=Sing|Person=3|Tense=Pres|VerbForm=Fin
4                                             Degree=Pos
5                                                   <NA>
6                              Mood=Imp|VerbForm=Fin
7 Case=Acc|Gender=Neut|Number=Sing|Person=3|PronType=Prs
8                                                   <NA>
  head_token_id  dep_rel deps         misc
1             2 compound <NA>         <NA>
2             4    nsubj <NA>         <NA>
3             4      cop <NA>         <NA>
4             0     root <NA> SpaceAfter=No
5             4    punct <NA>         <NA>
6             0     root <NA>         <NA>
7             1      obj <NA> SpaceAfter=No
8             1    punct <NA> SpacesAfter=\\n
```

```
> d <- c("Good product!",
+        "Bad product.",
+        "Not very good product.",
+        "Not bad.",
+        "Very very good.")
> d <- tolower(d)
> df <- document_term_frequencies(d)
```

```
> df
    doc_id    term freq
 1:   doc1    good    1
 2:   doc1 product    1
 3:   doc2     bad    1
 4:   doc2 product    1
 5:   doc3     not    1
 6:   doc3    very    1
 7:   doc3    good    1
 8:   doc3 product    1
 9:   doc4     not    1
10:   doc4     bad    1
11:   doc5    very    2
12:   doc5    good    1

> document_term_frequencies_statistics(df)
> df
    doc_id    term freq        tf       idf    tf_idf
 1:   doc1    good    1 0.5000000 0.5108256 0.2554128
 2:   doc1 product    1 0.5000000 0.5108256 0.2554128
 3:   doc2     bad    1 0.5000000 0.9162907 0.4581454
 4:   doc2 product    1 0.5000000 0.5108256 0.2554128
 5:   doc3     not    1 0.2500000 0.9162907 0.2290727
 6:   doc3    very    1 0.2500000 0.9162907 0.2290727
 7:   doc3    good    1 0.2500000 0.5108256 0.1277064
 8:   doc3 product    1 0.2500000 0.5108256 0.1277064
 9:   doc4     not    1 0.5000000 0.9162907 0.4581454
10:   doc4     bad    1 0.5000000 0.9162907 0.4581454
11:   doc5    very    2 0.6666667 0.9162907 0.6108605
12:   doc5    good    1 0.3333333 0.5108256 0.1702752
       tf_bm25      bm25
 1: 1.1042471 0.5640777
 2: 1.1042471 0.5640777
 3: 1.1042471 1.0118114
 4: 1.1042471 0.5640777
 5: 0.8194842 0.7508858
 6: 0.8194842 0.7508858
 7: 0.8194842 0.4186135
 8: 0.8194842 0.4186135
 9: 1.1042471 1.0118114
10: 1.1042471 1.0118114
11: 1.3179724 1.2076458
12: 0.9407895 0.4805794
```

```
> df <- document_term_frequencies(d)
> dtm <- document_term_matrix(df)
> dtm
5 x 5 sparse Matrix of class "dgCMatrix"
      bad good not product very
doc1   .   1   .        1    .
doc2   1   .   .        1    .
doc3   .   1   1        1    1
doc4   1   .   1        .    .
doc5   .   1   .        .    2

> dtm_colsums(dtm)
    bad      good      not product     very
      2         3        2       3        3
> dtm_rowsums(dtm)
doc1 doc2 doc3 doc4 doc5
   2    2    4    2    3

> dtm_remove_terms(dtm, terms = c("product"))
5 x 4 sparse Matrix of class "dgCMatrix"
      bad good not very
doc1   .   1   .    .
doc2   1   .   .    .
doc3   .   1   1    1
doc4   1   .   1    .
doc5   .   1   .    2

> as.matrix(dtm)
      bad good not product very
doc1   0   1   0        1    0
doc2   1   0   0        1    0
doc3   0   1   1        1    1
doc4   1   0   1        0    0
doc5   0   1   0        0    2
```

# Chapter 4

## Classification

## 4.1 Sample Data

The training and testing data come from a publicly available data set containing customer reviews of books purchased in Amazon e-shops. That this data can be freely downloaded from a publicly accessible URL [190]. Experiments, demonstrated below, use a randomly selected set from the large original database. The data collection for demonstrating capabilities of selected classifiers are limited here to just 1000 samples which are evenly split into two classes: books with positive reviews (five-star reviews) and books with negative reviews (one-star reviews). All reviews are written by customers in English.

To illustrate the original reviews, here are some randomly selected data items from the data set coming from the URL mentioned above [190] and used in this book as input for the presented classification algorithms. These demonstration reviews are not related to any particular book. To save space, long reviews have not been included here.

First, five positive reviews:

■ Really interesting. Everything you need, the most important is inside. Clear explanation, good pictures to represent 'the move'. This books helps me to improve myself and I will soon read it again... I guess it can be useful for beginner too...

■ This book is good on many levels. I learned about Ethiopia, it's culture and the struggles during this time. Excellent writing.

■ Ron Hansen is among the best writers delving into unusual real-life historical situations, and trying to make sense of them. Mariette is a pretty 17 year old drawn to the convent, who immediately becomes a special, vision influenced, faith inspired, stigmatized True Believer. Many of the older nuns are skeptical (to say the least), while the younger ones are in awe. Mr. Hansen guides us right into the heart and soul of this Catholic community, and of course Mariette. After reading this book, one feels a certain understanding of this potential saint, and how religion may become an obsession. I would personally agree with her physician / father, the scientific skeptic, but Mariette's story is certainly believable in the hands of a fine writer like Mr. Hansen!

■ Excellent book by a wonderful leader. Used it to develop a presentation on leadership for a mentoring team. Practical and excellent lessons for everyone. Thank you, Mr. Powell.

■ Harry Stein writes one of the most poltically honest memoirs I've read in a long time. In straightforward speak that skirts no liberl ideals, Stein chops down an ideology that he swore by in the 1960s but now sees as a hinderance to our modern-day culture. While his liberal foes cringe, every independently-minded American should give Stein's views a chance. For the naysers of this book, especially hardcore liberals, I found it interesting that many of the facts Stein used to back up his positions were pulled out of the liberally-slanted media he so accurately portrays.

Next, five negative reviews:

■ For people who have some knowledge of China, the book is of NO interest. Other books by James Mann (About Face), Jonathan Fenby, not to forget Spence (In Search of Modern China), are clearly superior.Kissinger's admiration for Mao is a disgrace. Mao was a murderer and Kissinger's repetitions admiration for China's civiliztion is non-sence. Over the centuries, China has been a mess, with short periods of success, but mostly calamities. Because of its size, China should have been a success, but it is not. Japan was must faster on its feet! China's isolation was a huge mistake. It is only recently, thanks to Deng and Western support, that China at last is emerging!Kissinger acknowledges that China has always played one "barbarian" against the other. It is very true but he does not draw any conclusion. To-day China is gaming both Europe and America: Just look at the "free" transfer of thechnology. The West is stupid and we should unite ton face China...The last point is Climate Change. To write about China and not discuss the catastrophic impact of climate change, is a major failure of the book. China - and Asia - will be destroyed by climate change.The combination of bad governance, corruption, and climate change is DEADLY. It is a Chinese and Asian disease.

■ Let me set the scene for you:I read the first three chapters of this book a few years ago right before it was published, and because I was immediately hooked and knew that I'd blow through the book in a day (then end up miserable until the next one was published over a year later), I decided to put it aside until there were more books to read. So fast forward to last week... I had an eight hour shift taking inventory at a local retailer scheduled, so I pulled up my Kindle wishlist and looked for something that was also available on Audible. This book was the winner, and–shame on me–I was, unfortunately, too excited to bother with listening to the Audible sample first. Oops. Big OOPS.Within thirty seconds I was asking myself if the producers of the audiobook were serious. By forty-five seconds, I was wondering if the author had approved the narrator and was happy with her. Then, around the sixty second mark, I decided it didn't matter to me. I turned it off. I couldn't listen to a story being read that way, no matter how compelling it is. So here I am; warning others away. READ the book–don't bother trying to listen to it. Or, if listening is your only option, then I would suggest looking into using Ivona Voices to create an MP3 version of the Kindle book. You would be better off, in my not so humble opinion .

■ Foolishly, I purchased a copy of this book for a young mother, at her request.Now, I am familiar with the contents of said book, and I don't know how to deal with the guilt. Oh, my God! What can I do to protect the daughter of this young woman? If she is naive enough to swallow this "author's" militant sickness, I have doomed an innocent child to a life of horror.On a more positive note, if I can convince the recipient to toss the volume into the dumpster, I will have re-moved one copy of it from circulation, and perhaps saved another innocent child from this abuse.This abomination is not recommended.

■ The Lombardi Rules attempts to provide a philosophical background of Coach Lombardi's beliefs and apply them to life and business. This was written by Vince Lombardi Jr, not the man himself, and although you'd expect his son to also pos-sess many of the inspiring attributes of Coach Lombardi, they are not exhibited in this book. It is essentially a collection of Vince Lombardi quotes turned into 2-3 page chapters. Making matters worse is that the text supporting the quotes ranges from cliche to painfully obvious with nothing profound or meaningful, not even a decent example from Lombardi's illustrious coaching career. Its al-ways frustrating when you could read the chapter headings and gain as much value as the entire book. You would be better served just searching for Vince Lombardi quotes than wasting any time with this book.

■ I normally adore anything and everything Victoria Alexander. This book started out to be good but I felt it quickly left me feeling bored and dissatisfied. The Princess's character was very annoying. Although I love Weston I just couldn't get passed how slow the book moved along. I would not recommend this book to a first time reader of Victoria Alexander. She is a wonderful writer, this book I feel was just a lemon out of hundreds.

Half of the examples used for the demonstration of classifiers, 500, are la-beled as positive reviews, the remaining half as negative. All reviews have been taken as they are, including possible grammar mistakes, typing errors, and so on, as typical for contributions on social networks. No special adjustment was done except for removing irrelevant attributes (those that do not contribute to classification). Removing such attributes decreases computational complexity (by lowering the number of dimensions, that is words, therefore saving memory and time). In the package tm, there is a very efficient function, removeSparseTerms(), reducing matrix sparsity (removing columns containing too many zeros according to parameters given by a user). The parameter was experimentally set to 0.995. This means that attributes that appear in at least 0.5% $(1 - 0.995)$ of documents are retained. This setting filtered a majority of attributes (only 2218 left, out of 17903) while retaining the classification accuracy.

The data are prepared in the pure textual form and their input format is CSV (comma separated values), including the first line which contains names of attributes (which are the words themselves). As for the format of the input data, it is represented simply as a vector for each review, where the values of elements are the word frequencies in individual reviews.

In a simplified way, vectors can be viewed like genomes composed of genes, where a genome constitutes a textual document and genes represent words. The more genes are shared between documents, the more similar ('related') the documents are. A vector direction points at a subspace containing similar documents.

The data representation is a matrix – every column represents an attribute, every row (vector) represents a textual review. The last vector element is a class label (which is a discrete value). In each of the following sections, the same demonstration data set is used to present all selected classifiers mentioned.

## 4.2 Selected Algorithms

In this book, a chosen set of algorithms consists of typical ones employed for text mining up to the present. Naturally, the machine-learning selected here algorithms do not cover every possibility and a reader is encouraged to experiment with many other existing methods. One of the problems of machine-learning is that there seems to be no universal, best, algorithm for any data type. In practice, it is often necessary to experimentally look for the best data/text mining tool.

This problem has been discussed in many publications; for example, refer to [281] and [282]. Finding and presenting the final optimal algorithm as a solution for a given application might sometimes just be that proverbial tip of the glacier, while the rest can absorb the overwhelming majority of invested time and other resources.

This part of the book presents machine-learning algorithms which employ various principles of learning:

■ Optimal and Naïve Bayes' Classifier – traditional and fast methods based on (conditional) probability theory, with excellent mathematical support, including uncertainty;

■ k-Nearest Neighbors – methods based on similarity with known entities, very fast during the training phase, might be appreciably slow during the classification process;

■ Decision Trees – popular 'divide and conquer' algorithms, which can provide understandable knowledge which may be transferred from a tree to a set of rules;

■ Random Forest – classification by a group of decision trees, which usually provides good results based on a noticeable reduction of possible tree correlations and over-fitting;

■ Support Vector Machines (SVM) – an often applied algorithm to problems where it is convenient to look for linear boundaries between classes in artificially enlarged spaces when the original space has just non-linear boundaries; SVM can give very good results but might have a high computational complexity;

■ Adaboost – a favorite method based on using a set of 'weak learners' that, together, can provide excellent results, even if each of the learners classifies with an error only slightly less than 50%.

■ Deep Learning – one of very modern methods employing the well-known and time-proven artificial neural networks in many applications. This approach relies on the power of multilayered networks, where the deepness is given by the number of those layers; each layer makes new combinations of input values, thus creating an abstract space where the classification can work well. Deep Learning is a resource (time, memory) demanding method, which can, today, take advantage of powerful hardware.

All the above mentioned algorithms have many implementations, including the popular language R. Sometimes, in the same programming language or a system, a user can select more or less different implementations according her or his application needs. One of the most important matters is to learn which parameters are available for an algorithm and which values it accepts, including the default values. The default algorithm parameters are often set to the statistically most used values; however, it does not have to be the optimal ones.

In this book, a reader cannot find a detailed description of all possible parameters because the intention was not to supply a manual. However, the detailed understandable descriptions, including samples of R-code, are available from the

Internet as PDF documents and may be easily downloaded from the following web pages:

■ Optimal and Naïve Bayes' Classifier – see [182],

■ k-Nearest Neighbors – see [226],

■ Decision Trees – see [154],

■ Random Trees and Forest – see [171],

■ Support Vector Machines (SVM) – see [153] and [193],

■ Adaboost – see [60],

■ Deep Learning – see [83].

## 4.3 Classifier Quality Measurement

After selecting and training a classifier algorithm, the supervisor needs to know the classifier efficiency – based on the training samples, is the mined knowledge sufficient enough for its successful application to future data?

Once the classifier training process has reached a certain peak value (depleted time, or maximum number of steps, or any other appropriate criterion), which completes the learning, it is necessary to express the algorithm's training quality, preferably numerically. This value can also be used to compare the mutual performance degree of various classifiers under investigation over the same data from a given application point of view.

Several methods are used in practice, of which the four most commonly used will be mentioned here. To evaluate the quality of a classifier, it is necessary to use so-called test data (samples), which correspond thematically to the training data, but were not used during training. It is similar to teaching at a school where a teacher (supervisor) teaches pupils to correctly compute some examples, and uses other, but similar examples to test their understanding.

From a binary perspective, four terms are used for the computation of classifier efficiency: *True Positive* (*TP*) for the number of data items classified correctly from a certain class $C$ point of view, *False Positive* (*FP*) for items incorrectly assigned to $C$, *True Negative* (*TN*) for items correctly classified as not belonging to $C$, and *False Negative* (*FN*) for items belonging to $C$ but assigned to other class or classes [107]:

■ *accuracy*: $(TP+TN)/(TP+TN+FP+FN)$, that is, the number of correct answers to the number of all classified samples, often expressed as a percentage;

■ *precision*: $TP/(TP+FP)$, that is, the number of items correctly labeled as belonging to the positive class $C$ divided by the total number of items labeled both correctly and incorrectly as belonging to $C$ (which is actually the number of relevant items in a set of items assigned to $C$ by the investigated classifier);

■ *recall*: $TP/(TP+FN)$, that is, the proportion of actual positives that was identified correctly; and

■ *F-measure*: it is the weighted harmonic mean of the precision and recall:

$$F = \frac{1}{\alpha \frac{1}{precision} + (1-\alpha)\frac{1}{recall}} = \frac{(\beta^2+1) \cdot precision \cdot recall}{\beta^2 \cdot precision + recall},$$

where $\beta^2 = \frac{1-\alpha}{\alpha}$. When $\beta$ increases, the precision has a higher importance. When the precision and recall are equally weighted (have the same importance), $\beta = 1$, and the measure is called also as $F1$ or *F-score*, which is the harmonic mean of the precision and recall:

$$F_1 = \frac{2 \cdot precision \cdot recall}{2 \cdot precision + recall} = \frac{2 \cdot TP}{2 \cdot TP + FP + FN}$$

It is possible to draw a graph where the horizontal axis represents the *recall* and the vertical axis represents the *precision*, however, sometimes not a curve but just a number, which expresses the classification quality, is needed.

Various metrics evaluating the results of classification can be calculated for each class. For example, the precision for class 0 is 0.5 and the precision for class 1 is 0.9. We often, we want just one number evaluating a classifier. Thus, the next step might be to average these two values. Using macroaveraging, which simply relies on the arithmetic mean, the average precision is 0.7. Both values representing a classifier's performance for each class have an equal weight. Microaveraging aggregates the decisions across all classes, taking every single classified item into consideration. Thus, the classes have different weights depending on their size (bigger classes contribute to a higher measure value). When class 0 is predicted for 20 items and class 1 for 100 items (class 1 is predicted five times more frequently than class 0), microaveraged precision will be $20/(20+100)*0.5+100/(20+100)*0.9 = 0.83$. It is obvious that the precision is than closer to the value of the precision for class 1, which is bigger.

# Chapter 5

# Bayes Classifier

## 5.1 Introduction

Bayes classification methods belong to the family of the oldest algorithms used in machine learning. Principal ideas of Bayes (or sometimes called Bayesian) classifiers are grounded in the works of a brilliant and famous English mathematician and statistician Thomas Bayes (1702, London - 1761, Tunbridge Wells, United Kingdom). He was the first who used probability inductively and established a broadly applied mathematical basis for probability inference – a means of calculating, from the number of times an event has occurred, the probability that it may again occur in future trials or observations.

One characteristic and important property of the probabilistic computations is that it works with *uncertainty* – probability very rarely returns an unequivocal answer to a question. Hence, a typical solution is the selection of the answer with the highest probability, taking into account that other possible answers might sometimes be correct as well, according to the support provided by their degree of probability.

Thomas Bayes' findings on probability were published posthumously in the *Philosophical Transactions of the Royal Society of London* as 'Essay Towards Solving a Problem in the Doctrine of Chances' (1763). His discovery, also known as Bayes' theorem, is directed at practically solving various real-world problems. The theorem not only addresses issues pertaining to Bayes' times but modern needs of data and information processing often take advantage of it as well often.

An interested reader may get acquainted with Thomas Bayes' original article by accessing this URL [19]. The article was communicated to John Canton, M. A. and F. R. S., in Newington Green, in a letter by Mr. Richard Price on November 10, 1763.

As an example related to text-mining, everyone who employs the commonly used electronic mail communication is also likely to use a kind of so-called spam filter, part of which is the implementation of a simplified form of Bayes' theorem widely known as *naïve Bayes classifier* (see more in Subsection 5.4 mentioned below). One of the big advantages of the Bayes classification methods is their excellent mathematical foundation in probability theory, particularly in conditional probability modeling.

Such models are generally applicable to any task where the knowledge represented by the probabilities of being a member of a certain class is acceptable, provided that a solved problem is described by training data supporting and enabling probability calculations. (A reminder that training data is a set of examples where the classification is known for each example.) Those probabilities can then play a role of weights, backing individual class memberships – a knowledge miner may select a hypothesis having the highest support.

It is necessary to keep in mind the fact that Bayes inference methods do not generally provide crisp, unequivocal answers of the type 'an unlabeled event $a_i$ belongs just to a class $C_j$ with the probability $p(a_i) = 1.0$ and certainly not to another class $C'_{k \neq j}$'. Usually, for such a probability $p(a_i)$, it holds true that $0.0 < p(a_i) < 1.0$, even if either $p(a_i) = 0.0$ or $p(a_i) = 1.0$ is also quite possible but not typical in practice.

## 5.2 Bayes' Theorem

Bayes' theorem is an interesting way to present, in more detail, the application of the probability theory to machine learning because this theory often plays, at least 'just' in the background, one of the fundamental roles in many algorithms. There are many good books that can be recommended to familiarize a reader with mathematical details, applications, and implementations of Bayesian methods. Deeper knowledge may be found, for example, in [23, 100, 160, 191, 129, 29] or many others available from assorted authors and publishers propounding various point of views, sometimes focused more on theory, at others more on practice.

The principle of Bayes' theorem can be simply explained using two events (sometimes called propositions) labeled $a$ and $b$, each having affiliated definitions:

- $p(a)$ ... an unconditional *a priori* probability of $a$ without any regard to $b$,

- $p(b)$ ... an unconditional *a priori* probability of $b$ without any regard to $a$,

- $p(a|b)$ ... a conditional probability of observing an event $a$ given that $b$ is true, i.e., $p(b) = 1.0$, and

■ $p(b|a)$ ... a conditional probability of observing an event $b$ given that $a$ is true, i.e., $p(a) = 1.0$,

where any probability $p(.)$ is defined within the real interval $0.0 \leq p(.) \leq 1.0$. The probabilities of events are typically computed from their observed frequencies – *a posteriori* probabilities, which is quite natural for real-world practical problems. Further, using the known *product rule* that both $a$ and $b$ must occur together while the order of their occurrence is insignificant, that is $a \wedge b = b \wedge a$, it is possible to write:

$$p(a \wedge b) \quad = \quad p(a|b)p(b), \tag{5.1}$$

$$p(b \wedge a) \quad = \quad p(b|a)p(a). \tag{5.2}$$

Note: $a$ and $b$ are two events that are generally *not* mutually independent.

Because the left sides of Eq. 5.1 and Eq. 5.2 describe the same event combination (logical conjunction $a$ *AND* $b$), it is possible to write the following equation using the right side equality of Eq. 5.1 and Eq. 5.2:

$$p(a|b)p(b) = p(b|a)p(a). \tag{5.3}$$

After dividing both sides of the formula Eq. 5.3 by $p(b)$, a new equation Eq. 5.4 can be acquired, which is that famous *Bayes' rule* or *Bayes' theorem*:

$$p(a|b) = \frac{p(b|a)p(a)}{p(b)}, \; p(b) > 0. \tag{5.4}$$

What does Eq. 5.4 actually say? The result, i.e. the enumerated left side $p(a|b)$, provides the *a posteriori* probability of $a$ given some observation of $b$.

The theory of probability can be used for extending the theorem to more than two events; however, the general computational complexity rapidly grows non-linearly, demanding the enumeration of many formula members. This is a disadvantage of Bayes' theorem, namely when it is applied to large spaces with a very high number of dimensions $x_i$ (which is typical for text mining where a term from a vocabulary represents a dimension).

Fortunately, there is a practical, even if not quite mathematically correct, solution known as naïve Bayes classifier mentioned in the following section – and it has been successfully used for a long time in many practical real-world areas, including text mining.

It is also possible to look at the problem from this perspective: Let $D$ stand for a set of collected data items (that is, all available observations up to a certain time), and $h_i \in H$ for an $i$-th hypothesis from a set $H$ of $n$ various hypotheses that are, to a certain degree, supported by that data, where $1 \leq i \leq n$.

The equation Eq. 5.5 can now be written in the following way:

$$p(h_i|D) = \frac{p(D|h_i)p(h_i)}{p(D)}, \ p(D) > 0 . \tag{5.5}$$

The question is: Which of the hypotheses has the highest support (that is, probability value) given that data $D$ is functioning as an empirical proof for a correct final decision? Bayes' theorem provides a way to calculate the *a posteriori* probability of each hypothesis $h_i$, usually referred to as a *maximum a posteriori* (MAP) hypothesis, $h_{MAP}$, which is defined using the following equation:

$$h_{MAP} = \underset{h \in H}{argmax} \ P(h|D) . \tag{5.6}$$

Using Eq. 5.5 and omitting $p(D)$ from its denominator (because it has the same value for each $h_i \in H$):

$$p(h_i|D) \propto p(D|h_i)p(h_i) . \tag{5.7}$$

Then, applying Eq. 5.7 to simplify Eq. 5.6, the latter can now be rewritten as a proportional relationship:

$$h_{MAP} \propto \underset{h \in H}{argmax} \ P(D|h)P(h) . \tag{5.8}$$

The most probable classification of a new, unlabeled instance can be obtained by combining the prediction of all hypotheses that are weighted by their posterior probabilities.

The question which arises is, how the previous equations can be used for inductive machine learning and supervised classification based on training using labeled samples.

In his book [199], the author Tom Mitchell demonstrates a nice example when $h_{MAP}$ may not provide the correct answer (suggesting 'a positive result'), which is supported by its *a posteriori* probability, let us say, 0.4 because other two hypotheses together suggest 'a negative result' (let us say, probabilities 0.3 and 0.3, i.e. together 0.6) and therefore their combined *a posteriori* probabilities outperform $h_{MAP}$.

## 5.3 Optimal Bayes Classifier

To determine the the strongest support of a hypothesis, it is necessary to combine the predictions of all hypotheses, weighted by their *a posteriori* probabilities. Let $C$ be a set of possible $m$ classes, $c_j$, $2 \le j \le m$, such that $c_j \in C$. The goal is to find the best classification $c_j$ supported by the observed data $D$, that is, having the highest conditional probability $p(c_j|D)$, which can be computed in this way:

$$p(c_j|D) = \sum_{h_i \in H} p(c_j|h_i)p(h_i|D) . \tag{5.9}$$

In such a case the most probable class-label assignment, $c_j$, to an unlabeled item is given by the maximum value of $p(c_j|D)$:

$$c_j = \underset{c_j \in C}{argmax} \sum_{h_i \in H} p(c_j|h_i)p(h_i|D) . \tag{5.10}$$

Note: *argmax* returns a class label, which is supported by the highest value of computed conditional probability.

## 5.4 Naïve Bayes Classifier

According to theory, the optimal Bayes classifier provides the best possible classification result based on given data and hypotheses. Unfortunately, in practice, the calculation – the sum of all multiplications – of all necessary combinations in Eq. 5.9 or Eq. 5.10 might be very demanding from the computational complexity point of view.

This is especially (but not only) true in text mining because assigning a correct class-label $c_j$ depends on the number of attributes, which here are words in social-media contributions or text documents. After collecting a big number of training samples (which is good from the probability computation point of vie), their vocabulary may be very large – thousands or tens of thousands unique words (or generally terms).

Each word has its probability in every class and there are very many possible attribute combinations, even if the data is reduced by, for example, eliminating insignificant terms.

Let $a_1, a_2, ..., a_N$ be the given attributes (vocabulary words). Then, using Eq. 5.8, it is possible to write for the best classification, $c_{MAP}$ :

$$c_{MAP} = \underset{c_j \in C}{argmax} \; p(a_1, a_2, \ldots, a_N|c_j)p(c_j) . \tag{5.11}$$

For large values of $N$'s, the computational complexity is given by the item $p(a_1, a_2, \ldots, a_N)$ in Eq. 5.11. Because of the possible mutual conditional dependence of attributes it is necessary to compute probabilities of all their possible combinations for a given $c_j$.

The idea of so-called *naïve Bayes classifier*, which is (theoretically not quite accurately) based on an assumption that there is no interdependence between the attributes (which is that 'naïvity') resulted as a way to make the computation easier and practically applicable. It means that, for the classification, it is possible to employ a simpler equation to determine $c_{MAP}$, which can be called as $c_{NB}$ (for *NB* like *Naïve Bayes*):

$$c_{NB} = \underset{c_j \in C}{argmax}\ p(c_j) \prod_i p(a_i | c_j) . \tag{5.12}$$

In other words, using the probabilities of attributes in the training data, just the product of *a posteriori* probabilities of observed attributes times *a priori* probability of a class occurrence per classified textual item has to be calculated. Due to that theoretical incorrectness, the classification result may be more frequently wrong, depending on how much the independence assumption is violated. More detailed explanations of the above-mentioned problems may be found in [199], which inspired the description of optimal Bayes' and naïve Bayes' classifier.

Despite the incorrectness, the naïve Bayes classifier is very popular because its results are quite acceptable in thousands of existing applications. The reason is that, in practice, usually only small fragments of attributes are somehow mutually dependent, if any. However, one should be watchful when interpreting the classification results – this incorrectness may be one of the reasons for the occasional wrong classification which users of e-mail know well: non-spam labeled as spam and vice versa.

## 5.5 Illustrative Example of Naïve Bayes

In conclusion, the following, absolutely simplified example with limited data just for demonstration, illustrates how the naïve Bayes' classifier works. Let the training data set be composed of seven sentences that express various descriptions of weather, a 'good/bad weather' problem.

Each sentence (a weather report) belongs to either a positive class $\oplus$ (good weather) or to a negative class $\ominus$ (bad weather) according to a certain categorization (let us say that *warm* is generally preferred to *cold*). This data set (using only lower case letters) is used for training the naïve Bayes' classifier:

1. it is nice weather $\oplus$
2. it is cold $\ominus$
3. it is not very cold $\oplus$
4. not nice $\ominus$
5. very cold $\ominus$
6. cold $\ominus$
7. very nice weather $\oplus$

The *csv* format of the training samples may, in a well arranged way with superfluous spaces, look like this (the numbers represent frequencies of words in documents):

```
it, is, nice, weather, cold, not, very, class
 1,  1,    1,       1,    0,   0,    0,   pos
 1,  1,    0,       0,    1,   0,    0,   neg
 1,  1,    0,       0,    1,   1,    1,   pos
 0,  0,    1,       0,    0,   1,    0,   neg
 0,  0,    0,       0,    1,   0,    1,   neg
 0,  0,    0,       0,    1,   0,    0,   neg
 0,  0,    1,       1,    0,   0,    1,   pos
```

Based on the knowledge obtained by generalization of the training samples, the classifier is expected to answer a question: Which label should be assigned to an unlabeled item *it is not nice cold at all* – a positive or negative one? What would a human being say? Probably *negative* because the proposition *it is not nice cold at all* does not sound as something positive from the semantic point of view, especially for people who like warm weather. However, what would be an answer from a computer equipped with a text-mining tool based on the naïve Bayes' classifier?

First, it is necessary to compute the required probabilities using the frequencies of words $w_i$, $i = 1, 2, \ldots, 7$, in the available training collection of samples:

$w_1 = it$: 2 occurrences in $\oplus$, 1 occurrence in $\ominus$, in total 3,
$w_2 = is$: 2 occurrences in $\oplus$, 1 occurrence in $\ominus$, in total 3,
$w_3 = nice$: 2 occurrences in $\oplus$, 1 occurrence in $\ominus$, in total 3,
$w_4 = weather$: 2 occurrences in $\oplus$, 0 occurrence in $\ominus$, in total 2,
$w_5 = cold$: 1 occurrences in $\oplus$, 3 occurrence in $\ominus$, in total 4,
$w_6 = not$: 1 occurrences in $\oplus$, 1 occurrence in $\ominus$, in total 2,
$w_7 = very$: 2 occurrences in $\oplus$, 1 occurrence in $\ominus$, in total 3.

The vocabulary contains 7 unique words (it, is, nice, weather, cold, not, very). In total, the training set consisting of 7 documents contains 20 words altogether (12 for $\oplus$ and 8 for $\ominus$); some words are repeated several times.

One problem can be seen: the analyzed item *it is not nice cold at all* includes two words *at* and *all*, which are not in the training set. In such a case, it is necessary to discard such words and work only with those that are in the training vocabulary.

Later, if the classification looks quite correct, the classifier can be retrained using the added labeled example(s) and the vocabulary can be correspondingly extended – such a procedure used to be applied in *semi-supervised learning* (progressively building a larger training set to improve a classifier's results).

Now, *at all* will be discharged here, leaving just *it is not nice cold* for the classification.

Based on words in the classified item and their statistical values from the vocabulary, provided that the occurrences of words are mutually independent (which can or cannot be quite correct), the naïve Bayes' classifier gives an answer: the decision of assigning either $\oplus$ or $\ominus$ to the unlabeled item is based on computing the probabilities $p_{NB_+}$ and $p_{NB_-}$, respectively, whether the item belongs to $\oplus$ or $\ominus$:

$$p_{NB_+}(\text{it, is, not, nice, cold} \mid \oplus) =$$
$$= p(\text{it} \mid \oplus)p(\text{is} \mid \oplus)p(\text{not} \mid \oplus)p(\text{nice} \mid \oplus)p(\text{cold} \mid \oplus)$$

and

$$p_{NB_-}(\text{it, is, not, nice, cold} \mid \ominus) =$$
$$= p(\text{it} \mid \ominus)p(\text{is} \mid \ominus)p(\text{not} \mid \ominus)p(\text{nice} \mid \ominus)p(\text{cold} \mid \ominus) .$$

The *a priori* probability of being in the class $\oplus$ is $p(\oplus) = 3/7 \doteq 0.429$ (3 positive sentences from a total of 7), and similarly in the class $\ominus$, $p(\ominus) = 4/7 \doteq 0.571$ (4 negative sentences from a total of 7).

Using all the known probabilities from the frequencies, everything is now ready for the computation of the classification supporting-weights $\wp_{NB}(\ominus)$ and $\wp_{NB}(\oplus)$ for each class $c_1 = \ominus$ and $c_2 = \oplus$, where $c_1, c_2 \in C = \{c_1 = \ominus, c_2 = \oplus\}$, as requested by the naïve Bayes' classifier according to Eq. 5.12:

$$\wp_{NB}(\oplus) = p(\oplus)p(\text{it}|\oplus)p(\text{is}|\oplus)p(\text{not}|\oplus)p(\text{nice}|\oplus)p(\text{cold}|\oplus) =$$
$$= 3/7 \cdot 2/12 \cdot 2/12 \cdot 1/12 \cdot 2/12 \cdot 1/12 \doteq$$
$$\doteq 0.429 \cdot 0.167 \cdot 0.167 \cdot 0.08 \cdot 0.167 \cdot 0.08 \cdot \doteq 0,00001279 ,$$

and

$$\wp_{NB}(\ominus) = p(\ominus)p(\text{it})|\ominus)p(\text{is}|\ominus)p(\text{not}|\ominus)p(\text{nice}|\ominus)p(\text{cold}|\ominus) =$$
$$= 4/7 \cdot 1/8 \cdot 1/8 \cdot 1/8 \cdot 1/8 \cdot 3/8 \doteq$$
$$\doteq 0.571 \cdot 0.125 \cdot 0.125 \cdot 0.125 \cdot 0.125 \cdot 0.375 \cdot \doteq 0,00005228 .$$

Thus, the result (class label) is:

$$c_{NB} = \underset{c_j \in C}{argmax} \left( \wp_{NB}(\ominus), \wp_{NB}(\oplus) \right) =$$
$$= \underset{\{c_1 = \ominus, c_2 = \oplus\}}{argmax} (5.228 \cdot 10^{-5},\ 1.279 \cdot 10^{-5}) = \ominus$$

because $\wp_{NB}(\ominus) > \wp_{NB}(\oplus)$ . The result is that the unlabeled item should obtain the label $\ominus$ and it belongs to the negative propositions, which corresponds with that human expectation briefly discussed above.

Note that the results $\wp(.)$'s do not directly provide probability values.

They are numerical values playing the role of weights and as such may be safely used for selecting a hypothesis $c_{NB}$: $p(.) \neq \wp(.)$, however, $p(.) \propto \wp(.)$.

Should it be – for any reason – necessary to express $\wp_{NB}(.)$'s as probabilities $p_{NB}(.)$'s, it is possible to normalize both output numbers $\wp_{NB}(.)$ within the closed probability interval $[0.0, 1.0]$:

$$p_{NB}(\ominus) = \frac{\wp_{NB}(\ominus)}{\wp_{NB}(\oplus) + \wp_{NB}(\ominus)} \text{ and } p_{NB}(\oplus) = \frac{\wp_{NB}(\oplus)}{\wp_{NB}(\oplus) + \wp_{NB}(\ominus)}.$$

After substituting appropriate numbers for symbols, the result is expressed with the requested probabilities. In this case, the probability of being *bad weather* is approximately four times higher than that of being *good weather*, thus the result is well supported and quite acceptable:

$$p_{NB}(\ominus) = 0.00005228 \,/\, (0.00005228 + 0.00001279) \doteq 0.8034,$$
$$p_{NB}(\oplus) = 0.00001279 \,/\, (0.00005228 + 0.00001279) \doteq 0.1966,$$
$$p_{NB}(\ominus) + p_{NB}(\oplus) = 0.8034 + 0.1966 = 1.0.$$

As a reader may see, there are, in fact, two possible results that can be true, however, neither of them is fully true because neither $p_{NB}(\ominus)$ nor $p_{NB}(\oplus)$ are equal to 1.0. Therefore, the result is subject to uncertainty and a user has to decide which value is worth of consideration. In reality, the result with the highest probability is typically selected which is reasonable; however a user can expect that such a system would not always work without occasional errors, even if the output would mostly be correct. If $p_{NB}(\ominus) = p_{NB}(\oplus)$, then the class can be selected randomly (the highest uncertainty, an indecisive case).

People who use electronic mail with a spam filter know it well: sometimes, a message is not spam but is considered to be, and vice versa. The reason is that the positive and negative class usually share many words (terms, phrases) like in the example presented above – and the bag-of-words representation, which uses separated individual words without their mutual relationships, introduces part of that uncertainty. A word *bad* (or *nice*) can be in both the positive and the negative class and even the combination of their probabilities does not entirely remove the uncertainty. Sometimes it could also be difficult for human beings – as a reader is likely to know.

In conclusion, one technical note. When computing the resulting probability by multiplication of many individual probabilities, there is a danger of numerical underflow, especially for classification of text documents using a broad vocabulary because the average probability value of a word is very low. That is why such data-mining tools, instead of working with multiplication, employ summarization of logarithms of probability values.

## 5.6 Naïve Bayes Classifier in R

The system R offers several implementations of the naïve Bayes classifier. An implementation which will be installed into R library, `naivebayes`, is briefly described in this section along with an example demonstrating how to use it for text mining.

## 5.6.1   Running Naïve Bayes Classifier in RStudio

For simplicity, let us use one directory for data (in which there is for demonstration, the input file, *BookReviews1000.csv*, containing 1000 samples for training and testing) as well as the R-code for running the naïve Bayes classifier.

In the first step, RStudio is started and the source R-code is open. After clicking on the RStudio menu button *Session*, the selection of *Set Working Directory* and finally *To Source Location* defines the working directory, where source inputs are found and where prospective outputs will be stored.

*Console* confirms this selection. For example:

```
> setwd("D:/R Book/Naive Bayes")
```

The source code should run on clicking the button, *Source*.

If the package `naivebayes` has never been employed, there may be a red error message:

```
Error in library (naivebayes) : there is no package
called 'naivebayes'
```

This is likely because the R library (which is one of the directories where R is installed) does not implicitly contain such a package. In this case, it is necessary to install `naivebayes`, which is quite simple as is illustrated in Figure 5.1:

1. click *Packages*
2. click *Install*
3. write the package name *naivebayes* into the pop up window
4. select *Install dependencies* (if not selected by default)
5. click *Install* in the pop up window

That is all that is required; the R-system will automatically install everything necessary. The list of all installed libraries can be seen in RStudio under the *System Library* label. After successful installation, the naïve Bayes classifier can run as described as follows (text lines can be wrapped when they are longer than the available page width, and the wrapped part are indented – console messages and R-code including comments – but actually they create just one line!).

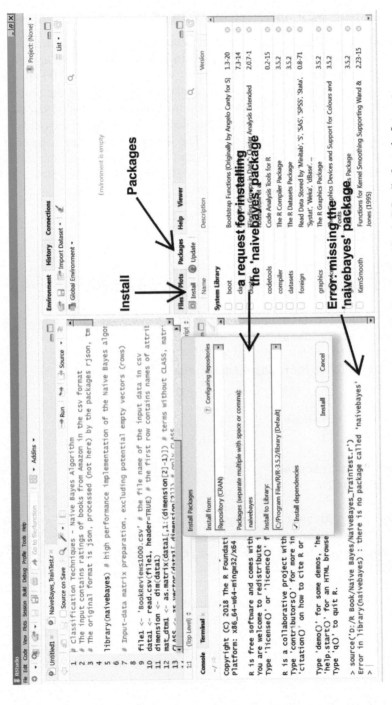

**Figure 5.1:** Installing the package, *naivebayes*, for the naïve Bayes classifier in RStudio from the Internet.

## 5.6.2    Testing with an External Dataset

This section illustrates possible R-code for generating and testing a trained naïve Bayes classifier using an extra test data-set. In this case, the simple approach is applied just for the demonstration. In a matrix with 1000 samples, 100 randomly selected positive and negative samples are taken for testing, and the remaining 900 for training. Each row in the matrix contains a randomly selected sample, the sequence of which is also random from the class point of view, so it would not be necessary to resample it randomly again. The selected 100 testing rows contain 48 positive samples and 52 negative ones, and the remaining, training samples, 452 positive and 448 negative ones, which is almost the ideal ratio, 50% : 50%.

Here is the sample R-code:

```
library(naivebayes)

# loading the data
data <- read.csv("BookReviews.csv", header=TRUE)

# converting the data to a matrix
dtm <- as.matrix(data)

# removing the object data (unnecessary from now)
rm(data)

# number of instances
n <- dim(dtm)[1]

# number of attributes
m <- dim(dtm)[2]

# indexes of 100 testing samples
number_of_test_samples <- 100
test_index <- sample(n, number_of_test_samples)

# selecting training samples
train_data <- dtm[-test_index, 1:(m-1)]
train_labels <- dtm[-test_index, m]
# selecting testing samples
test_data <- dtm[test_index, 1:(m-1)]
test_labels <- dtm[test_index, m]

# training a Naive Bayes model
nb_model <- naive_bayes(x=train_data, y=train_labels)
```

```r
# calculating class predictions
predictions <- predict(nb_model, test_data, type="class")

# creating a confusion matrix
cm <- table(test_labels, predictions)
print("Confusion matrix")
print(cm)

# number of instances
n <- sum(cm)

# number of correctly classified instances for each class
correct <- diag(cm)

# numbers of instances in each class
instances_in_classes <- apply(cm, 1, sum)

# numbers of each class predictions
class_predictions <- apply(cm, 2, sum)

# accuracy
accuracy <- sum(correct)/n

# precision per class
precision <- correct/class_predictions

# recall per class
recall <- correct/instances_in_classes

# F1-measure per class
f1 <- 2 * precision * recall / (precision + recall)

# printing summary information for all classes
df <- data.frame(precision, recall, f1)
print("Detailed classification metrics")
print(df)
print(paste("Accuracy:", accuracy))

# macroaveraging
print("Macro-averaged metrics")
print(colMeans(df))
```

```
# microaveraging
print("Micro-averaged metrics")
print(apply(df, 2, function (x)
                    weighted.mean(x, w=instances_in_classes)))
```

The console output after a successful run:

```
[1] "Confusion matrix"
           predictions
test_labels  0  1
           0 38 14
           1  9 39
[1] "Detailed classification metrics"
   precision     recall         f1
0 0.8085106 0.7307692 0.7676768
1 0.7358491 0.8125000 0.7722772
[1] "Accuracy: 0.77"

[1] "Macro-averaged metrics"
precision      recall         f1
0.7721798 0.7716346 0.7699770

[1] "Micro-averaged metrics"
precision      recall         f1
0.7736331 0.7700000 0.7698850
```

## 5.6.3   Testing with 10-Fold Cross-Validation

As another example for the same data, the code uses a standard 10-fold cross-validation to gradually test the classifier accuracy.

There are 10 steps, each one uses 10% of the samples for testing and 90% for training, and every sample gradually takes part in both sections. Classification performance metrics are calculated at each step. Their final value is given by the mean value of all 10 measurements. In our example, only the values of accuracy are considered for simplicity. The standard error of the mean (SEM) that is used in statistics for expressing an average error for different randomly selected testing samples from a data population (an error generalization for the whole population) is also calculated.

Given below is the sample R-code. A reader would notice that the code for cross-validation testing is very similar to the code using a test set for testing.

In fact, the same steps (selecting the data, building a model, using the model, evaluating the model) are repeated ten times. In each iteration, different training and test sets are created. Classification performance measures are calculated in each iteration and the results are averaged at the end. To prevent repeating an almost identical code in each chapter related to classification, the R code for cross-validation testing is not included in the following chapters.

```
library(naivebayes)

# a function for calculating the standard error of the mean
sem <- function(x) sd(x)/sqrt(length(x))

# loading the data
data <- read.csv("BookReviews.csv", header=TRUE)

# converting the data to a matrix
dtm <- as.matrix(data)

# removing the object data (unnecessary from now)
rm(data)

# number of instances
n <- dim(dtm)[1]

# number of attributes
m <- dim(dtm)[2]

# 10 folds for cross-validation
number_of_folds <- 10

# marking samples in each of 10 folds with 1's, 2's, ..., 10's
folds <- cut(1:n, breaks=number_of_folds, labels=FALSE)

# vector with accuracies
accuracies <- rep(0, times=number_of_folds)

for(i in 1:number_of_folds){
    # selecting testing samples according to their fold-marks
    test_index <- which(folds==i, arr.ind=TRUE)

    # selecting training samples
    train_data <- dtm[-test_index, 1:(m-1)]
    train_labels <- dtm[-test_index, m]
```

```r
    # selecting testing samples
    test_data <- dtm[test_index, 1:(m-1)]
    test_labels <- dtm[test_index, m]

    # training a Naive Bayes model
    nb_model <- naive_bayes(x=train_data, y=train_labels)

    # calculating class predictions
    predictions <- predict(nb_model, test_data, type="class")

    # obtaining a confusion matrix
    cm <- table(test_labels, predictions)

    # number of correctly classified instances for each class
    correct <- diag(cm)

    # numbers of each class predictions
    class_predictions <- apply(cm, 2, sum)

    # accuracy
    accuracy <- sum(correct)/sum(cm)

    # precision per class
    precision <- correct/class_predictions

    # printing the results
    cat("Cross-validation fold: ", i, "\n",
        " Accuracy: ", accuracy, "\n",
        " Precision(positive): ", precision["1"], "\n",
        " Precision(negative): ", precision["0"], "\n",
        sep="")

    # storing the accuracy for current crossvalidation fold
    accuracies[i] <- accuracy
}
# printing the aggregated results from crossvalidation
cat("Average Accuracy: ",
    round(sum(accuracies)/number_of_folds*100, 1), "%\n",
    "SEM: ", round(sem(accuracies)*100, 2),
    sep="")
```

The console output after the successful run:

```
Cross-validation fold:  1
 Accuracy:  0.75
 Precision(positive):  0.6666667
 Precision(negative):  0.8478261
Cross-validation fold:  2
 Accuracy:  0.67
 Precision(positive):  0.6792453
 Precision(negative):  0.6595745
Cross-validation fold:  3
 Accuracy:  0.68
 Precision(positive):  0.6521739
 Precision(negative):  0.7419355
Cross-validation fold:  4
 Accuracy:  0.75
 Precision(positive):  0.6909091
 Precision(negative):  0.8222222
Cross-validation fold:  5
 Accuracy:  0.67
 Precision(positive):  0.6323529
 Precision(negative):  0.75
Cross-validation fold:  6
 Accuracy:  0.67
 Precision(positive):  0.6470588
 Precision(negative):  0.71875
Cross-validation fold:  7
 Accuracy:  0.64
 Precision(positive):  0.5409836
 Precision(negative):  0.7948718
Cross-validation fold:  8
 Accuracy:  0.67
 Precision(positive):  0.6515152
 Precision(negative):  0.7058824
Cross-validation fold:  9
 Accuracy:  0.64
 Precision(positive):  0.6229508
 Precision(negative):  0.6666667
```

```
Cross-validation fold:    10
  Accuracy:   0.66
  Precision(positive):    0.6716418
  Precision(negative):    0.6363636
  Average Accuracy: 68%
SEM: 1.24
```

What are such results good for? The following point of view can be considered, for example, to summarize and somehow interpret the results. Each testing subset for each of ten folds is classified using the particular naïve Bayes model created by available training samples. The standard error of the mean (SEM), averaging classification errors for all individual random selections (each such selection generally a has different error mean), is only 1.24%. It is a good sign when the SEM is small because of stability – it suggests that the created model does not depend (too much) on different data samples and is stable.

Looking at individual fold results, it is obvious that negative examples have lower classification errors than their positive counterparts (the same is valid also for testing by the test data subset). This could be an interesting discovery, a kind of dug out knowledge from the data, and an inspiration for subsequent research (which is not included in this book). Why is it so?

A possible explanation might be that – after a closer exploration of words and phrases in both data classes – the positive reviews are, for example, mutually more different (readers like more things) while the negative ones are more homogeneous in terms of what readers dislike. In such a case, the generalization of data (which might be the requested knowledge) may be more difficult for the positive section and 'positive' readers might be separated into several 'positive', more homogeneous, groups.

On the other hand, if there were many more (let us say, 95%) training negative reviews than positive ones (an unbalanced training set) – the conclusion might be that a book is generally not very well accepted and its reprint would not pay off. Or, that publishing of the next volume is not promising because all the previous volumes did not demonstrate non-declining interest, and so on.

# Chapter 6

# Nearest Neighbors

## 6.1 Introduction

Unlike the Bayes' rule, the family of algorithms called $k$-NN ($k$ Nearest Neighbors) does not use the probability theory. It also ranks among time-tested methods, successful (as well as sometimes regrettably unsuccessful) in many applications.

The $k$-NN algorithm is based on using *similarity* – the principal idea is that similar items belong to the same class. Therefore, it is necessary just to collect and store the convenient number of labeled samples belonging to each relevant class – and the training phase is completed. This is undoubtedly a big advantage of the $k$-NN algorithm: very fast and simple training procedure. In addition, it is a reason why $k$-NN is counted among the so-called *lazy algorithms*. A good discussion about $k$-NN can be found in [76].

During the classification process, $k$-NN waits for an incoming unlabeled item, finds its $k \geq 1$ most similar labeled sample(s) and assigns the appropriate label to the unlabeled item. Unlike the training phase, the classification one is typically much more demanding and often computationally complex due to the time-consuming similarity determination of a classified item in relation to every stored sample. More samples naturally means more information; however, also more computation including its non-linearly increasing complexity, which depends on the sample number as well as on the dimensionality extent.

## 6.2 Similarity as Distance

If an unlabeled item $x_i$ is the most similar (or identical) to a certain labeled sample $s_j$, then $x_i$ is labeled just as $s_j$: $s_j \in C_l$, $x_i \sim s_j \rightarrow x_i \in C_l$, or in other words, $C_l \subset C_l \cap x_i$, where $C_l$ is one of classes appropriate to the analyzed problem. Thus. the element $s_j$ is called the *nearest neighbor* of $x_i$.

This is, basically a very simple algorithm; however, the question is, how to compute that similarity.? Several methods of the numeric expression of similarity, which are based on computing a kind of distance between pairs of points, which represent individual samples in an $n$-dimensional space may be used. Depending on the properties of the space under consideration, there are numerous methods for calculating the distance.

*Euclidean distance*: The commonly known and habitually used distance named after the 'founder of geometry,' the ancient Greek mathematician Eukleides (Euclid) of Alexandria (born c. 325 BC, died c. 270 BC). This distance is the straight-line distance between two points in Euclidean space, which is the two-dimensional plane and three-dimensional space, i.e., it can be the $n$-dimensional hyperplane using the Cartesian coordinate system. To compute the Euclidean distance $d_E$ between a point $A$ with $n$ coordinates $a_1, a_2, \ldots, a_n$, and a point $B$ with coordinates $b_1, b_2, \ldots, b_n$, the familiar formula is applied:

$$d_E(A, B) = \sqrt{(a_1 - b_1)^2 + (a_2 - b_2)^2 + \ldots + (a_n - b_n)^2} = \sqrt{\sum_{i=1}^{n} (a_i - b_i)^2} \quad (6.1)$$

*Manhattan distance*: The distance between two points is the sum of the absolute differences of their Cartesian coordinates. This distance $d_M$, named after the layout of most streets on the island of Manhattan (a borough of New York City), is an alternative to the Euclidean distance when it is possible to move in the space only in parallel with mutually perpendicular axes:

$$d_M(A, B) = |a_1 - b_1| + |a_2 - b_2| + \ldots + |a_n - b_n| = \sum_{i=1}^{n} |a_i - b_i| \quad (6.2)$$

*Hamming distance*: A very simple distance between two strings (or vectors) of equal length, which is the number of positions at which the corresponding symbols are either equal or different. The more different the symbols, the less similar (or more distant) the strings. The Hamming distance, named after a mathematician Richard Wesley Hamming (11th February 1915 - 7th January 1998), can also be used to determine the similarity or difference between two text documents when the representation of words is binary (for example, using symbols 1 and 0, for when a word occurs = 1, and a word does not occur = 0) provided that all such represented documents employ the same joint vocabulary (which guarantees the required same length of vector).

*Mahalanobis distance*: Named after an Indian statistician, Prasanta Chandra Mahalanobis ($29^{th}$ June 1893 - $28^{th}$ June 1972), the distance used in statistics. It measures a point distance from a mean of a distribution using covariance matrix. It might be useful when there is correlation between points representing training documents.

*Cosine distance*: This similarity can be used in text mining when documents with very different number of words are compared, in which case the scalar frequency representation (coordinates of a point) may not provide reliable results; for example, an abstract of an article and the article itself, or a short review compared to a long one expressing the same opinion. In such a case, each document is taken as a non-zero vector pointing from the origin of the coordinate system to its endpoint given by the coordinates and the mutual similarity is measured by the angle between pairs of the vectors. If the angle $\alpha$ between two vectors is zero, they are considered to be identical because $\cos(0°) = 1.0$ (or 100% similarity). If $\alpha = 90°$, then $\cos(90°) = 0.0$ (or complete dissimilarity). Otherwise, for $0° < \alpha < 90°$, the similarity is something in between. Note that any distance or frequency cannot be negative, therefore no negative cosine values are used.

The cosine similarity for two vectors $\mathbf{x}_j$ and $\mathbf{x}_k$ can be written as the following formula:

$$\cos(\alpha) = \frac{\mathbf{x}_j \cdot \mathbf{x}_k}{||\mathbf{x}_j||\,||\mathbf{x}_k||} = \frac{\sum_{i=1}^{n} x_{ji} \cdot x_{ki}}{\sqrt{\sum_{i=1}^{n} x_{ji}^2}\,\sqrt{\sum_{i=1}^{n} x_{ki}^2}} \tag{6.3}$$

Figure 6.1 illustrates a very simple case. There are three documents $doc_1$, $doc_2$, and $doc_3$. The vocabulary space is limited to just two words $w_i$ and $w_j$ represented by their frequencies in the documents, $f_{w_i}$ and $f_{w_j}$, respectively. The documents $doc_1$ and $doc_2$ have different absolute numbers of the two word

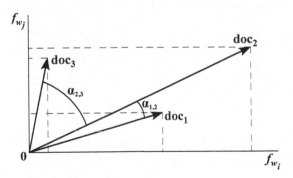

**Figure 6.1:** The illustration of the cosine distance (similarity) between documents represented by vectors. A simplified two-dimensional space with frequencies of $word_i$ and $word_j$.

frequencies, however, their rate $f_{w_i}:f_{w_j}$ is almost the same for both documents. From the Euclidean point of view, the documents are far apart, hence different.

On the contrary, from the cosine similarity point of view, they are very close, because the $\alpha_{1,2}$ angle between them is considerably small. The document $doc_3$ contains only a small number of words $w_i$ and, in addition, the $doc_3$ frequency rate $f_{w_i}:f_{w_j}$ differs from both $doc_1$ and $doc_2$ rates which is also the reason for the big angle $\alpha_{2,3}$ and significant dissimilarity apart from the Euclidean viewpoint.

An interested reader can find much deeper description of various distance/similarity methods including their details in, for example, [156] and [212].

## 6.3 Illustrative Example of $k$-NN

The small and simple data set, which consists of seven textual documents related to the weather description (introduced earlier in the section 5.5 Illustrative Example of Naïve Bayes) can illustrate the application of the $k$-NN algorithm to classification. The frequencies of the seven vocabulary words in the individual documents act as coordinates of the documents in a seven-dimensional abstract space (seven attributes/variables $w_1 = $ it, $w_2 = $ is, ..., $w_7 = $ very). Each of the seven training documents is represented as a point in that space labeled by its class name, either *pos* or *neg*.

The classified document position in the space is known, too, thus it is necessary to find the nearest point, which is labeled, and to then assign that label to the unlabeled item.

The training data-set with an added sequence number, No.:

```
No. it, is, nice, weather, cold, not, very, class
1.  1,  1,  1,      1,      0,   0,   0,    pos
2.  1,  1,  0,      0,      1,   0,   0,    neg
3.  1,  1,  0,      0,      1,   1,   1,    pos
4.  0,  0,  1,      0,      0,   1,   0,    neg
5.  0,  0,  0,      0,      1,   0,   1,    neg
6.  0,  0,  0,      0,      1,   0,   0,    neg
7.  0,  0,  1,      1,      0,   0,   1,    pos
```

The unlabeled document calling for its classification (the mark '?' stands here for an unknown label) looks like this:

```
1,  1,   1,       0,    1,   1,    0,    ?
```

Provided that there is no reason to avoid the Euclidean space, the distances of the classified point from the labeled samples can be computed and then the nearest sample's label may be used. Due to the simplicity of the training data-set, the frequencies here are only 0's or 1's, but that does not change anything.

For the first training document, its Euclidean distance $d_1$ from the classified point is:

$$\sqrt{(1-1)^2+(1-1)^2+(1-1)^2+(0-1)^2+(1-0)^2+(1-0)^2+(0-0)^2} =$$
$$= \sqrt{3} \doteq 1.73 \, .$$

Similarly, all the six remaining distances (from the $2^{nd}$ up to the $7^{th}$ training sample) are calculated as well: $d_2 = 1.41$, $d_3 = 1.41$, $d_4 = 1.73$, $d_5 = 2.24$, $d_6 = 2.0$, and $d_7 = 2.45$.

Evidently, the shortest distance is 1.41 – unfortunately, it is shared by two training neighbors (the $2^{nd}$ and $3^{rd}$), each one from a different class.

So, 1-NN does not work (nor 2-NN), it is an irresolute case. What about 3-NN? This is again a draw, there are two candidates for it (the $1^{st}$ and $4^{th}$) having the same distance 1.73. We could consider 5-NN. Adding the sample No. 6 with its distance 2.0 and label *neg*, there are now two *positive* and three *negative* samples; therefore, the voting majority finally decides: the classified item gets the label, *neg*.

As there are only word frequencies like 0 and 1, it is possible to employ the computationally simpler binary representation together with the Hamming distance.

Evidently, the distances, $d$, between the unlabeled example and the samples No. 1 to 7, are 3, 2, 2, 3, 5, 4, and 6, respectively. The closest samples are No. 2 ($d = 2$. *neg*) and No. 3 ($d = 2$, *pos*) so the result is a draw. After adding the next closest neighbors No. 1 (*pos*) and No. 4 (*neg*) with $d = 3$, that is, a draw again. The closest neighbor which follows is No. 6 (*neg*) with $d = 4$, thus the rate is 3 *neg* : 2 *pos*, so the assigned label is *neg*.

This illustration shows a possible disadvantage of $k$-NN, namely the difficulty in determining the number $k$ of neighbors in advance. In practice, a researcher often has to determine the right number of nearest neighbors experimentally.

In any case, at the beginning, *all* the distances between the unlabeled point and the points with labels must be calculated, they must then be ordered by size, and finally, used for the decision.

If the close neighbourhood for small $k$'s gives poor results, the surroundings of an unlabeled point should be enlarged (increasing $k$). It is, however, necessary to be cautious if just a narrow majority prevails because, for example, a result 50:51 (for $k=101$) is less compelling than 2:1 ($k=3$) or 3:2 or 4:1 ($k=5$), even if it may be correct and acceptable – it depends on a particular application and its data type. When in doubt, the results of alternative algorithms might be taken in account, even if only for comparison.

The correct nearest neighbor selection may be affected by grossly different scale ratios on the axes. If justified reasons do not preclude it, it is advisable to normalize the scales on the axes to, for example, the interval [0, 1]:

$$\xi_i = \frac{x_i - min(x_i)}{max(x_i) - min(x_i)},$$

where $\xi_i$ is a normalized value of $x_i$, $min(x_i)$ is the minimum value and $max(x_i)$ the maximum value on a given axis obtained from the training samples. The illustration, Figure 6.2, shows how a classified case (marked with '?') can change its classification from a class B to a class A due to normalization of scales. The need for normalization can, in case of doubt, be confirmed or rejected by the testing procedure.

It is also worth mentioning here other possible problems concerning (not only) $k$-NN, namely the influence of a very high-dimensional space on classification results, where, for example, the influence of a relevant attribute can be overridden by a large number of irrelevant ones, as seen in [96].

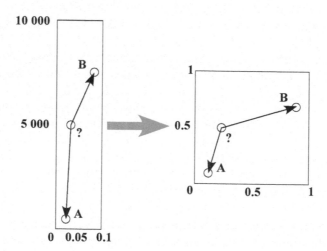

**Figure 6.2:** Very different scale ratios (on the left side) of different dimensions may cause the nearest neighbor to be incorrectly identified. After normalization (on the right side), the situation may be quite the opposite.

## 6.4   $k$-NN in R

The two examples which follow demonstrate the simple application of the $k$-NN algorithm to the same data, *BookReviews.csv*, as in the previous chapters. The R classification library called *class* and just one neighbor ($k = 1$) are used. Alternatively, the reader may simply try to change the parameter *k* to see if a higher number of neighbors will provide better classification results (accuracy). The first sample uses a test set:

```
library(class)

# loading the data
data <- read.csv('BookReviews.csv', header=TRUE)

# converting the data to a matrix
dtm <- as.matrix(data)

# removing the object data (unnecessary from now)
rm(data)

# number of instances
n <- dim(dtm)[1]

# number of attributes
m <- dim(dtm)[2]

# indexes of 100 testing samples
number_of_test_samples <- 100
test_index <- 1:number_of_test_samples

# selecting training samples
train_data <- dtm[-test_index, 1:(m-1)]
train_labels <- dtm[-test_index, m]

# selecting testing samples
test_data <- dtm[test_index, 1:(m-1)]
test_labels <- dtm[test_index, m]

# training a k-nearest neighbors model
knn_model <- knn(train_data, test_data, train_labels, k=1)

# creating a confusion matrix
cm <- table(test_labels, knn_model)
print("Confusion matrix")
print(cm)
```

```
# number of instances
n <- sum(cm)

# number of correctly classified instances for each class
correct <- diag(cm)

# numbers of instances in each class
instances_in_classes <- apply(cm, 1, sum)

# numbers of each class predictions
class_predictions <- apply(cm, 2, sum)

# accuracy
accuracy <- sum(correct)/n

# precision per class
precision <- correct/class_predictions

# recall per class
recall <- correct/instances_in_classes

# F1-measure per class
f1 <- 2 * precision * recall / (precision + recall)

# printing summary information for all classes
df <- data.frame(precision, recall, f1)
print("Detailed classification metrics")
print(df)
print(paste("Accuracy:", accuracy))

# macroaveraging
print("Macro-averaged metrics")
print(colMeans(df))

# microaveraging
print("Micro-averaged metrics")
print(apply(df, 2, function (x)
                 weighted.mean(x, w=instances_in_classes)))
```

The console output after a successful run:

```
[1] "Confusion matrix"
          knn_model
test_labels  0  1
          0 35 22
          1 15 28

[1] "Detailed classification metrics"
  precision    recall        f1
0      0.70 0.6140351 0.6542056
1      0.56 0.6511628 0.6021505
[1] "Accuracy: 0.63"

[1] "Macro-averaged metrics"
precision    recall        f1
0.6300000 0.6325989 0.6281781

[1] "Micro-averaged metrics"
precision    recall        f1
0.6398000 0.6300000 0.6318219
```

The output from *k*-NN classification using the cross-validation testing fol-lows:

```
Cross-validation fold:  1
 Accuracy:  0.64
 Precision(positive):  0.5714286
 Precision(negative):  0.7058824
Cross-validation fold:  2
 Accuracy:  0.59
 Precision(positive):  0.6222222
 Precision(negative):  0.5636364
Cross-validation fold:  3
 Accuracy:  0.56
 Precision(positive):  0.5957447
 Precision(negative):  0.5283019
 Cross-validation fold:  4
 Accuracy:  0.55
 Precision(positive):  0.5102041
 Precision(negative):  0.5882353
```

```
Cross-validation fold:  5
 Accuracy:  0.56
 Precision(positive):  0.5686275
 Precision(negative):  0.5510204
Cross-validation fold:  6
 Accuracy:  0.58
 Precision(positive):  0.5901639
 Precision(negative):  0.5641026
Cross-validation fold:  7
 Accuracy:  0.58
 Precision(positive):  0.4878049
 Precision(negative):  0.6440678
Cross-validation fold:  8
 Accuracy:  0.62
 Precision(positive):  0.6470588
 Precision(negative):  0.5918367
Cross-validation fold:  9
 Accuracy:  0.5
 Precision(positive):  0.5111111
 Precision(negative):  0.4909091
Cross-validation fold:  10
 Accuracy:  0.51
 Precision(positive):  0.6052632
 Precision(negative):  0.4516129
Average Accuracy:  56.9 %
SEM:  1.38
```

The reader can see the rather different results between the two different training and testing methods demonstrated above. This can be expected because the training and testing examples are not the same, indicating, among other things, that the selected classifier is not robust enough for the data. In such cases, a deeper analysis can be recommended to find answers to questions as to whether the training data is sufficiently representative, if the amount of data is sufficient, or if the selected classification algorithm is appropriate for the given task and should not it be replaced by a more suitable one.

# Chapter 7

# Decision Trees

## 7.1 Introduction

Generally speaking, *decision trees* represent a certain type of graph known as trees, which are acyclic connected graphs. Details can be found in the mathematics branch *graph theory*; see for example, [238, 105]. Trees are commonly used in computer science and informatics for various purposes, to represent a hierarchical data structure, for example.

One of the popular traditional machine learning tools is a graph called *rooted directed tree*, in which the edges are directed from a node called *root*. An edge can point either to another (sub)tree or a node called *leaf*.

The path from the root to a leaf creates a *branch*. Every leaf is the terminal node of its path. A tree can generally be composed of any number of other trees that are called *subtrees*, thus modeling a hierarchy.

The properties of the rooted directed tree may be applied to creating models known as *divide et impera* (divide and rule), which is one of the reasons why these trees belong among the more popular machine learning algorithms for classification tasks. Another reason is that decision trees (if not excessively extensive) provide relatively understandable representation of knowledge, where a branch can also be taken as a rule. Leaves provide the answers to questions given by combinations of attribute values – in each node, there is a certain test for a specific value of an attribute and the tree branches from that node, each branch either to another node (subtree) on a lower level or to a terminal node called a leaf. If the tree method is used for classification, each leaf contains a label of a class. and the classified item receives its class according to specific combinations of attribute values, which are decided in the node.

Generally, to solve a problem, there can often be more than one alternative tree, and it is even possible that all such trees may provide the same result. Having several models available, an application should use the model with the lowest classification error. If there are trees with the same error (or without an error), then the simplest one is selected, following the practical *Occam's razor rule*: Simpler explanations are more likely to be correct; avoid unnecessary or improbable assumptions. A simpler tree means a tree with a lower number of nodes.

If a tree with a lower (zero) error is available, it must be used. Is there such a tree? A reader may try to create other possible trees (starting with different attribute tests in the tree root) to discover if there is one better than the suggested simple tree, which could be an interesting task.

In reality, it is not always possible to generate a tree that classifies with the zero error. The problem might be insufficient training data, non-linear boundaries between classes (a tree separates the problem space into subspaces using hyperplanes parallel with attribute axes), or something else.

In such a case, an application must settle for imperfection or look for a better solution by testing alternate algorithms. On the other hand, when a decision tree is a suitable solution for a given application and its data, the practice life prefers simpler solutions according to the Occam's razor rule. It means looking for the simplest possible tree which solves a given classification problem. In machine learning, several possible solutions have been found; for an example, see [36]. This chapter is going to deal with the very widely used tree-generating method known as *c5-tree*, which employs a concept for the minimization of trees called *entropy minimization*.

## 7.2    Entropy Minimization-Based *c5* Algorithm

In a nutshell, the *entropy* concept in informatics works with so-called *chaos*, which means that a set may have elements belonging to different classes. The goal is to split the set into subsets where each subset contains elements of just one class – then the entropy of such a subset is zero, or there is no *chaos*. The maximum entropy is in a set where mixed elements of various classes are evenly represented by the same number of their occurrence.

### 7.2.1    The Principle of Generating Trees

Sorting out such mixed up elements via entropy-minimization was the idea of John Ross Quinlan, who also wrote an interesting book which is a classic today, on the same [220]. J. R. Quinlan is also an author of the first academic (and as well as commercial) successful implementation of his classification algorithm originally named *ID3* (for discrete attributes) and later, after extending it for discrete and continuous attributes, missing attribute values, pruning, and some other improvements, *c4.5* (one half of his book is a code in C). In the course of time, this was continually modified, developing into an effective software tool now known as *c5*; see also [224]. *C5* is also available as a single-threaded version (not only) in R (GNU license) and its use is demonstrated in the text which follows.

Entropy is defined with the help of probabilities. It was introduced by one of the 'fathers' of informatics, Claude Elwood Shannon (1916-2001); for details, see [249]. According to Shannon, when there are two possibilities having probabilities, $p$ and $1 - p$, the entropy, $H$, is defined as follows: Let $X$ be a discrete random variable that can take values from the set $\{x_1, x_2, \ldots, x_n\}$. Then $H(X)$ defines the entropy of $X$ this way:

$$H(X) = -\sum_{i=1}^{n} p(x_i) \log_2 p(x_i) \tag{7.1}$$

The logarithm base 2 is here because the information is measured in bits. To put it simply, at least $m_b$ bits are needed for $m$ registered cases stored in a set, so $m = 2^{m_b}$, therefore $m_b = log_2(m)$. If the probability, $p$, of selecting any case from the set is evenly distributed, then $p = 1/m$, therefore $m_b = log_2(1/p) = -log_2(p)$. In addition, the minus sign enables working with just positive numbers $(0.0 \le H(X) \le 1.0)$ because numerically negative information is neither defined nor used in informatics.

When a set is purely homogeneous, the probability of having elements of only one class is 1 and for other classes 0. Thus, it is guaranteed that a randomly selected element from such a set belongs to a certain class with probability $p = 1$ and, selecting an element from a different class has the probability $p = 0$. Therefore, $p \cdot \log_2 p = 1 \cdot \log_2 1 = 0$, which is the minimum entropy, and (for two classes) if the ratio of elements in that set is 50% : 50%, that is, $p = 0.5$, then the entropy is maximum because $0 \cdot \log_2 0 = 1$ (after processing by L'Hôpital's rule). In other words, if the entropy is 0, it is not necessary to investigate the contents of a set because the information about it is maximum.

The graph, Figure 7.1 illustrates the entropy curve shape expressing Eq. 7.1. Consider the horizontal axis $X$, a homogeneous set has the probability $p = 0$ that it contains an element which does not belong to it – therefore, the entropy of that set is minimum, $H(X) = 0$. Similarly, on the other end of the axis, $p = 1$, that is, any selected element from that set belongs only to that set – the set is also quite homogeneous and its $H(X) = 0$. If the set contains elements belonging to other sets, it is heterogeneous and the worse case is when $p = 0.5$, or there is a mixture of 'own' and 'foreign' elements with the ratio 50% : 50% for two classes: $H(X) = 1$, maximum entropy.

According to Quinlan's idea, a generated decision tree's task is not only to divide the original heterogeneous set into more homogeneous subsets but also reduce its size with respect to the number of nodes to as small as possible. Theoretically, it is possible to generate all possible trees and then choose the optimal

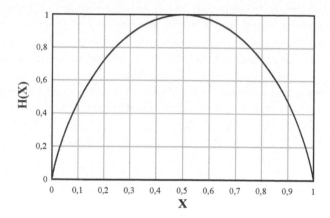

**Figure 7.1:** Entropy $H(X)$ curve shape.

one (lowest in error and size). Practically, however, it cannot be done because there may be too many trees and the computational complexity would be too high. In artificial intelligence, including machine learning, such cases can use some heuristics, while these do not guarantee optimal achievement, the result may be quite acceptable.

It should be noted here that the attributes of samples are supposed to be mutually independent – if the assumption is not fulfilled (which may for example, be due to significant correlation), the results may not be reliable. Therefore, attributes that depend on others should be discarded before processing. Such an assumption is, in machine learning, typical.

The *C5* tree generator successively tries one attribute after another to find which one decreases the entropy the most. Such an attribute $a_i$ is then selected as a node, which – in accordance with the results of testing its value – creates a new, hierarchically lower level in the tree, a subtree.

This tree node $a_i$ divides the heterogeneous set into more homogeneous subsets for the given training date set $T$ so that the eventual average entropy value of the resulting subsets maximally reduces the original entropy $H(T)$ of their parent set.

Therefore, the selected attribute $a_i$ provides the highest *information gain* $= H(T, a_i) = H(T) - H(T|a_i)$ – in other words, the maximal entropy difference, or decrease.

The average entropy is usually calculated using the following formula based on *a posteriori* probabilities and the individual entropies of newly formed leaves:

$$H_{avg} = \sum_b \frac{n_b}{n_t} \cdot (\text{chaos in a branch } b) = \sum_b \frac{n_b}{n_t} \cdot \sum_c \left( -\frac{n_{b.c}}{n_b} \log_2 \frac{n_{b.c}}{n_b} \right), \quad (7.2)$$

where $n_b$ is the number of samples in a branch $b$, $n_t$ the number of samples in all branches leading from the splitting note, and $n_{b,c}$ the total number of all samples in the branch $b$ of a class $c$.

As a reader can see, the formula 7.2 calculates the average entropy of a sub-tree (created by dividing a heterogeneous set into more homogeneous subsets) by calculating the entropy for each subset (the second sum on the right side) and weighting it by the relative number of samples it covers (the first sum on the right side). Note this is not the only possible formula but it is, in practice, commonly used due to good results it obtains.

If it is not possible to further reduce the entropy, the tree generating process stops and the branches end in leaves. Otherwise, the whole process continues recursively, testing all attributes again and again. It cannot be excluded that such a tree is not theoretically optimal; practically, however, this method provides very good outputs and this is why it is so favored in research and applications.

To demonstrate the contribution of an attribute to the entropy decrease, let us take again a simple decision problem 'good/bad weather' demonstrated already in section 5.5 :

1. it is nice weather $\oplus$
2. it is cold $\ominus$
3. it is not very cold $\oplus$
4. not nice $\ominus$
5. very cold $\ominus$
6. cold $\ominus$
7. very nice weather $\oplus$

In the original set of training samples, members of both classes (three positive, *pos*, $\oplus$, and four negative, *neg*, $\ominus$) are mixed together, hence the chaos.

The entropy of this considerably heterogeneous set can be quantified using Eq. 7.1 as follows (recall that $0 \cdot \log_b 0 = 0$, and $\log_b x = \log_{10} x / \log_{10} b$, thus $\log_2 x = \log_{10} x / \log_{10} 2$, or instead of $\log_{10}$ $\log_e$ can be used as well):

$$H(X) = -\frac{3}{7} \cdot \log_2 \frac{3}{7} - \frac{4}{7} \cdot \log_2 \frac{4}{7} \doteq 0.9853 \; .$$

As it can be seen, the entropy is very high, close to 1.0 because the ratio of the number of members in both classes is 3 $\oplus$ : 4 $\ominus$. For example, using the attribute *weather* (with frequencies either 0 or 1 in a sample) for splitting the original set into two more homogeneous subsets, and taking the help of Eq. 7.2 for computing the average entropy $H_{avg}$ after splitting, a very simple decision tree generated by c5 could look like the graph in Figure 7.2. The division assumes that if the word 'weather' is included, it is a positive case, otherwise a negative one.

This approach correctly covers two positive cases (No. 1 and 7); however, one case (No. 3) is classified incorrectly. All the negative cases are covered correctly.

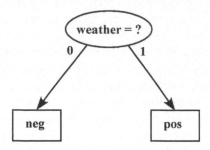

**Figure 7.2:** A decision tree solving the problem 'good/bad weather' with one error produced by the left branch ($H > 0$): without *weather* it is a negative message (one incorrect classification), otherwise a positive one (the right branch with $H = 0$).

The contribution of *weather* to the entropy decrease may, therefore, be numerically calculated as follows:

$$H_{avg} = \frac{5}{7} \cdot \left( -\frac{4}{5} \log_2 \frac{4}{5} - \frac{1}{5} \log_2 \frac{1}{5} \right) + \frac{2}{7} \cdot \left( -\frac{2}{2} \log_2 \frac{2}{2} - \frac{0}{2} \log_2 \frac{0}{2} \right) \doteq 0.5157 .$$

The calculation computes the left branch entropy (which is $> 0$) and the right branch entropy (which is $= 0$) for each class, employing the respective *a posteriori* probabilities in the expressions $p \cdot \log_2 p$ as well as the weights (5 cases for the left and 2 cases for the right branch out of 7).

Obviously, the initial entropy was significantly reduced and the not negligible information gain is $0.9853 - 0.5157 = 0.4696$. An interested reader may similarly compute the entropies, contributions of other attributes, and/or add some additional splitting to further decrease the chaos.

## 7.2.2   Pruning

The generated tree, without any further modifications, can simply serve as the exact model of the specific data from which it was created. However, if the tree is to be used for future classifications or predictions, it should represent the most general knowledge derived from the particular training data. In such a case, the so-called *pruning* process is used to improve the generalization of the obtained model. Pruning means the replacement of a subtree by a leaf, that is. a certain simplification of the tree.

From the point of view of training examples, this simplification may lead to a higher classification error related to *training* examples, which the unpruned tree can otherwise correctly classify into classes even optimally, with a zero error. Such a tree is so-called *over-fitted*, that is, it can properly classify mainly or only the training examples and fails to correctly process others. However, the aim is to also best classify cases that are not (or cannot be) included in the training set.

Practice shows that some generalization of the tree by pruning can increase (sometimes significantly) the classification accuracy on future data at the cost of a modest reduction of the same for the training data. The tree pruning rate can be controlled by a parameter which indicates the maximum allowed reduction of the tree. Typically, it is limited to 25 percent but sometimes, it is worth experimentally finding an optimal value.

The idea of the pruning process is illustrated in Figure 7.3. On the left side, a subtree perfectly assigns classes (white and black points). The node $x_j$ has a left branch directed to a leaf covering five white cases, which is fairly general in comparison with the right-directing branch leading to a leaf covering only one, thus, too specific case – a black point.

Replacing the whole subtree $x_j$ (consisting of the node $x_j$ and its two leaves) with the left-sided leaf of $x_i$ (see the right side of the figure) containing five correct white and one incorrect black point slightly increases the classification error measured by the training samples. However, it may decrease the error measured by testing samples – the over-fitting is reduced in favor of generality.

As an example, let us consider two possible solutions of the weather problem discussed above (see Figure 7.2 and Figure 7.4). The weather-tree in Figure 7.2 says that if the word *weather* is present in a text example, the decision is *pos*: good weather; otherwise, *neg*: bad weather. According to the generated tree, no other word is necessary for the decision: no *good, bad, nice, cold,* etc. Is it a correct decision tree? Should it look like that? Can there be a better, more accurate tree? As for the correctness, the presented decision tree in Figure 7.2 makes one mistake because in the sample 3 (which is correctly *pos*) attribute *weather* is not present but the tree classifies it wrongly as *neg*. The remaining six samples are classified correctly, so the classification accuracy is approximately 85.7%.

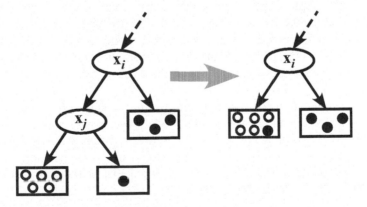

**Figure 7.3:** Replacing a subtree (on the left) with a leaf (on the right) by pruning.

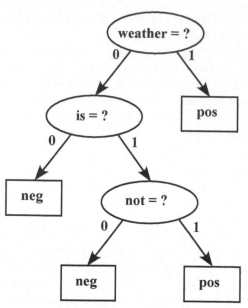

**Figure 7.4:** A full-size, unpruned decision tree variant dealing perfectly with the content meaning problem of the weather phrases 'good/bad weather'.

As is apparent, the model is not quite perfect, but from the point of view of generality, could that one mistake be forgiven – or not? This can only be evaluated on the basis of testing, which was not done here because of the triviality of the artificial demonstration task and the absence of training samples.

Figure 7.4 shows an alternative tree for the same training data. However, this time it is a complete variant which classifies the training data perfectly, without that one error in the previous case as demonstrated in Figure 7.2. The tree in Figure 7.2 may represent the pruned tree shown in Figure 7.4, replacing two testing nodes and three leaves with just one leaf for the price of a single mistake. Again, it is not possible to assess the generality of both trees without testing. It is obvious, however, that (in view of the new circumstances) the tree in Figure 7.4 is more complex, testing values of three attributes (*weather, is, not*), not just one (*weather*).

From the application point of view, the question is, is it worthwhile to use the more complicated tree or just the simpler model? The answer must be sought by users themselves, based on the needs of the produced application including time and memory demands. Models in the real world are usually much more extensive and more complicated; thus, it is very important to test them as carefully as possible before they are put into practice.

# 7.3 C5 Tree Generator in R

Using the same dataset, *BookReviews1000.csv*, with samples of book reviews, this section presents a possible application of the *C5* tree classifier to text mining.

First, the creation of a decision tree is demonstrated. Second, the tree properties are briefly discussed, pointing out the acquired useful information. Third, how to assess the accuracy of tree classification is demonstrated so that it can be applied to future data. Finally, the ability of the tree to provide understandable knowledge in the form of rules is shown.

## 7.3.1 Generating a Tree

In the first demonstration, the first 100 samples serve as testing items while the remaining 900 are used for training. The source R-code, employing the C50 package, used for generating the *C5*-tree including the corresponding testing rules and using both testing samples is as follows:

```
library(C50)

# loading the data
data <- read.csv("BookReviews.csv", header=TRUE)

# converting the data to a matrix
dtm <- as.matrix(data)

# removing the object data (unnecessary from now)
rm(data)

# number of instances
n <- dim(dtm)[1]

# number of attributes
m <- dim(dtm)[2]

# indexes of 100 testing samples
number_of_test_samples <- 100
test_index <- 1:number_of_test_samples


# selecting training samples
train_data <- dtm[-test_index, 1:(m-1)]
train_labels <- dtm[-test_index, m]
```

```
# selecting testing samples
test_data <- dtm[test_index, 1:(m-1)]
test_labels <- dtm[test_index, m]

# training a c5 model (tree and rules)
c5_tree <- C5.0(train_data, as.factor(train_labels))
c5_rules <- C5.0(train_data, as.factor(train_labels),
                 rules=TRUE)

# visualizing the generated decision tree
plot(c5_tree)

# printing the detailed summary for the generated
# decision tree and rules
print(summary(c5_tree))
print(summary(c5_rules))

# calculating class predictions for the tree
predictions <- predict(c5_tree, test_data, type="class")

# creating a confusion matrix
cm <- table(test_labels, predictions)
print("Confusion matrix")
print(cm)

# number of instances
n <- sum(cm)

# number of correctly classified instances for each class
correct <- diag(cm)

# numbers of instances in each class
instances_in_classes <- apply(cm, 1, sum)

# numbers of each class predictions
class_predictions <- apply(cm, 2, sum)

# accuracy
accuracy <- sum(correct)/n

# precision per class
precision <- correct/class_predictions
```

```
# recall per class
recall <- correct/instances_in_classes

# F1-measure per class
f1 <- 2 * precision * recall / (precision + recall)

# printing summary information for all classes
df <- data.frame(precision, recall, f1)
print("Detailed classification metrics")
print(df)
print(paste("Accuracy:", accuracy))

# macroaveraging
print("Macro-averaged metrics")
print(colMeans(df))

# microaveraging
print("Micro-averaged metrics")
print(apply(df, 2, function (x)
               weighted.mean(x, w=instances_in_classes)))
```

The *C5* tool generates a tree model according to the training samples. The tree is then displayed together with the result describing the classification of training samples. The console output is a result in a text-like form, where the positive class is marked as 1 and the negative one as 0. The tree is not trivial, so it is divided between several parts (subtrees). Due to the tree size, it is not copied here in total, just the root and a couple of levels under it have been shown:

```
Call:
C5.0.default(x = train_data, y = as.factor(train_labels))


C5.0 [Release 2.07 GPL Edition]      Tue Feb 19 09:03:50 2019
-------------------------------

Class specified by attribute 'outcome'

Read 900 cases (2218 attributes) from undefined.data

Decision tree:

waste > 0:
```

```
:...more <= 1: 0 (40)
:    more > 1: 1 (3/1)
waste <= 0:
:...boring > 0:
    :...day <= 0: 0 (49/1)
    :    day > 0: 1 (4/1)
    boring <= 0:
    :...tale > 0:
        :...apparently <= 0: 1 (25/1)
        :    apparently > 0: 0 (3/1)
        tale <= 0:
```

The rest of this tree (after the branch `tale <= 0`) has been left out. Its graphical depiction is shown in Figure 7.5 (again, only the top of the tree).

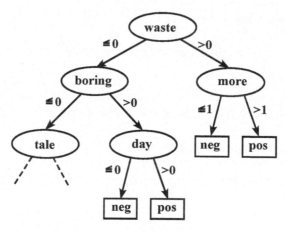

**Figure 7.5:** The generated C5-tree (only its beginning and the root attribute *waste* are shown here).

## 7.3.2 Information Acquired from C5-Tree

The word *waste* is present in the root. From the entropy minimization point of view, it is the best attribute and each classification starts with a test checking whether the numeric representation of this word (in this case, its frequency in a document) is $> 0$ or $\leq 0$ (however, frequency cannot be $< 0$).

After dividing the documents (the original set) between two categories (subsets) with and without *waste*, the entropy can be further reduced because, on the next level, the word *more* again divides documents (those without *waste*) between two categories. This branch ends because further lowering of the entropy is not possible here.

The branch leading to the class *0* covers exactly 40 documents having the frequency of the word *more* ≤ 1, while the other branch covers 4 documents containing *more* having its frequency > 1; however, one of them is classified incorrectly (in the *C5* output, it is indicated by 3/1, or 3 correctly classified and 1 incorrectly).

Documents without *waste* can be further divided into the category with *boring* and without *boring,* and then again using *tale* and *day*, and so on. Documents having no *boring* but with the attribute *tale* may be further divided to reduce the entropy, which is not illustrated at this point (but can be seen in the complete computational output), while the word *day* finishes a branch: without *day*, the classification is 0 (*negative*) for 49/1 cases, otherwise 1 (*positive*) for 4/1. Individual branches make rules together; for example, if *waste* = 0 ∧ *boring* > 0 ∧ *day* > 0 then *positive.*

In the textual tree representation below, the summary says that the tree size is 108 nodes. From the classification point of view, the accuracy error measured on training data is 4.9%, that is 44 from 900 samples. In addition, the appended confusion matrix shows how many negative samples (marked as 0) were classified correctly (425) and incorrectly (18). Similarly for the positive samples: 431 correctly, 26 incorrectly. There are, altogether, (18 + 26) − 44 incorrect classifications. Optimally, the confusion matrix should contain only non-zero numbers in its main diagonal (everything correctly classified) while the minor diagonals should contain only zeros (for any number of classes):

```
Evaluation on training data (900 cases):

        Decision Tree
        ----------------
        Size        Errors

        108    44( 4.9%)    <<


        (a)    (b)     <-classified as
        ----   ----
        425    18      (a): class 0
         26   431      (b): class 1
```

From the text-mining point of view, the generated tree also provides valuable information concerning so-called relevant attributes. The relevancy here means how much an attribute contributes to entropy reduction, that is, revealing what is typically common for each class – which words or phrases. A relevant attribute efficiently separates cases belonging to different classes. A list of relevant attributes is available (under `Attribute usage` heading) in the tree generator output. The attributes are ordered according to the number (in percentage) of times they are used for classification. An attribute in the tree root is tested every time (100%). Attributes on lower levels are used less and less. Many attributes (often a very large majority) are never used, thus they are insignificant from the classification point of view; here, it is only 96 from 2218, or 4.33%. In the experiment presented, the tree gives such a list of relevant attributes; see Figure 7.6.

The activity of the relevant attributes on different tree levels can be studied in the complete computational output; for example, which values are used for branching or how significant they are.

| 100.00% | waste | 13.67% | been | 5.44% | out | 2.89% | own |
|---|---|---|---|---|---|---|---|
| 95.22% | boring | 13.44% | her | 5.33% | job | 2.78% | control |
| 89.33% | tale | 12.89% | when | 5.22% | title | 2.78% | ridiculous |
| 86.22% | finish | 12.67% | believe | 4.78% | each | 2.67% | loved |
| 82.33% | poor | 12.56% | about | 4.78% | more | 2.67% | things |
| 79.22% | dont | 12.33% | for. | 4.33% | best | 2.56% | the |
| 68.78% | wonderful | 12.00% | without | 4.11% | attention | 2.44% | looking |
| 65.33% | authors | 10.67% | personal | 3.89% | feel | 1.67% | that |
| 61.67% | pages | 10.44% | theme | 3.78% | everything | 1.56% | can |
| 56.89% | young | 9.67% | while | 3.67% | giving | 1.56% | much |
| 54.00% | off | 9.44% | one | 3.67% | questions | 1.44% | has |
| 50.56% | life | 9.22% | before | 3.56% | book | 1.33% | and |
| 47.89% | not | 9.00% | true | 3.44% | buy | 1.33% | books |
| 44.22% | over | 8.22% | classic | 3.44% | hoping | 1.33% | enough |
| 31.67% | put | 8.11% | two | 3.33% | along | 1.33% | gives |
| 27.67% | only | 7.78% | took | 3.22% | couple | 1.22% | like |
| 27.56% | great | 7.33% | using | 3.11% | apparently | 1.11% | could |
| 26.11% | see | 7.22% | possible | 3.11% | beginning | 1.11% | nothing |
| 24.67% | after | 7.00% | think | 3.11% | brings | 1.00% | half |
| 23.00% | enjoyed | 6.56% | will | 3.11% | came | 0.89% | action |
| 20.56% | into | 6.33% | library | 3.00% | favorite | 0.78% | author |
| 15.22% | also | 6.00% | piece | 3.00% | less | 0.67% | trilogy |
| 14.44% | anyone | 5.89% | day | 3.00% | love | 0.56% | its |
| 14.00% | excellent | 5.67% | someone | 2.89% | hollywood | 0.44% | because |

**Figure 7.6:** The list of 96 relevant attributes generated by the c5 decision tree. The attributes are ordered according to their frequency of their use. 100% (*waste*) in the root is the input attribute tested always.

## 7.3.3 Using Testing Samples to Assess Tree Accuracy

The low classification error, 4.9%, indicates that the tree-model itself is very good, with a high accuracy 95.1%. If the goal is to create a model based on just the available data, without the intention to use it for the classification/prediction applied to new data in the future, a data/text miner might be fully satisfied. On the other hand, such a model represents only data from the past and can be strongly tied to the used training dataset – overtrained, without sufficient generalization that could be applied to unseen data items, whether from the past or the future.

This is the reason why the expected classification accuracy for the future data must be estimated using a disjunctive set of samples – testing samples. In the R-code above, the initial 100 samples were used for testing because those items were not used for training. The model created with the mentioned 900 training samples gives the following output on RStudio console:

```
[1] "Confusion matrix"
            predictions
test_labels  0  1
          0 36 21
          1 14 29

[1] "Detailed classification metrics"
   precision    recall         f1
0       0.72 0.6315789 0.6728972
1       0.58 0.6744186 0.6236559
[1] "Accuracy: 0.65"

[1] "Macro-averaged metrics"
precision      recall         f1
0.6500000 0.6529988 0.6482766

[1] "Micro-averaged metrics"
precision      recall         f1
0.6598000 0.6500000 0.6517234
```

In this case, the accuracy is appreciable lower, just 65%. Both positive and negative review samples have a very similar precision, 65% and 72%, respectively. The overall error, 35%, is relatively high, indicating that the model is not very good for the classification of unseen data.

A possible explanation might be that the training samples did not sufficiently cover the investigated area because only 900 reviews related to various book topics was not enough. It is likely the main reason as there are thousands of topics and millions of book readers.

A remedy for this situation could be to collect larger training data in subsets, which would cover more specific topics. Then, a model for each such subset could be generated, creating not only one universal classifier but a set of 'experts' for individual topics – which is how people do it. In any case, such situations need careful analysis. An alternative approach might also be to try several different classification algorithms in the hope that one of them would provide better results for a given application.

## 7.3.4   Using Cross-Validation to Assess Tree Accuracy

Secondly, the testing process applies the 10-fold cross-validation method. The code is not presented here, since it is identical to the cross-validation code in Chapter 5, except for two rows, creating and using a decision tree model instead of a probabilistic model. The results for each of 10 computation steps as well as for the final summarizing step that provides the overall accuracy are available on Console:

```
Cross-validation fold:   1
  Accuracy:   0.65
  Precision(positive):   0.58
  Precision(negative):   0.72
Cross-validation fold:   2
  Accuracy:   0.74
  Precision(positive):   0.76
  Precision(negative):   0.72
Cross-validation fold:   3
  Accuracy:   0.62
  Precision(positive):   0.6530612
  Precision(negative):   0.5882353
Cross-validation fold:   4
  Accuracy:   0.63
  Precision(positive):   0.5849057
  Precision(negative):   0.6808511
Cross-validation fold:   5
  Accuracy:   0.63
  Precision(positive):   0.6346154
  Precision(negative):   0.625
Cross-validation fold:   6
  Accuracy:   0.64
  Precision(positive):   0.6808511
  Precision(negative):   0.6037736
```

```
Cross-validation fold:   7
 Accuracy:   0.58
 Precision(positive):   0.4901961
 Precision(negative):   0.6734694
Cross-validation fold:   8
 Accuracy:   0.59
 Precision(positive):   0.62
 Precision(negative):   0.56
Cross-validation fold:   9
 Accuracy:   0.53
 Precision(positive):   0.537037
 Precision(negative):   0.5217391
Cross-validation fold:   10
 Accuracy:   0.74
 Precision(positive):   0.7818182
 Precision(negative):   0.6888889
Average Accuracy:   63.5 %
SEM:   2.07
```

Apparently, the result is very similar to the testing with 100 external samples. It confirms also, the mean error value = 63.5%, as well as a low standard error of the mean, SEM = 2.07, which evidences that there is small dispersion of results for each of ten steps. Therefore, the model is stable; however, its classification accuracy is not very good.

The conclusion can be the same as in the case with the external 100 testing samples briefly discussed above.

### 7.3.5   Generating Decision Rules

Decision rules, having the format *IF-THEN,* are generally favored because they represent relatively easily understandable knowledge, and if there is a 'reasonably' sized set of rules (for example, ten), their application is preferred – provided that the rules work well with high accuracy, given their application area. Rules can also be used, for example, for filling up a knowledge base.

The principle of generating rules consists of taking individual tree branches from the root to leaves; each branch may represent a rule. This procedure can also involve certain optimization like removing redundant rules; for example, in the case of the rules *IF $x_1$ THEN y* and *IF $x_1 \wedge x_2$ THEN y*, only the first rule suffices.

The C5-tree generator was applied to the demonstration data and the task was to create an adequate set of rules. In all, 20 rules were generated and the output shows them all. Only the first two (as copied from the output) are shown here for the purpose of illustration. Each rule represents a standard conjunction where attributes are tested for their frequency like in the tree nodes, and a leaf is on the right side of a rule. The transcription of the first rule into a more compact form looks like this (*dont* comes from *don't*, class 0 is *negative*):

IF *before* $\leq 0 \wedge$ *classic* $\leq 0 \wedge$ *dont* $> 0 \wedge$ *finish* $\leq 0 \wedge$ *not* $> 0 \wedge$ *possible* $\leq 0 \wedge$ *put* $\leq 0 \wedge$ *theme* $\leq 0 \wedge$ *took* $\leq 0$ *THEN class* $= 0$

A piece of the *c5* output:

```
Rules:

Rule 1:  (44, lift 2.0)
         before <= 0
         classic <= 0
         dont > 0
         finish <= 0
         not > 0
         possible <= 0
         put <= 0
         theme <= 0
         took <= 0
         ->  class 0   [0.978]

Rule 2:  (26, lift 2.0)
         along <= 0
         authors > 0
         best <= 0
         brings <= 0
         dont <= 0
         giving <= 0
         own <= 0
         wonderful <= 0
         ->  class 0   [0.964]
```

In the first rule, (44, lift 2.0) means 44 covered samples without an error (otherwise it is the same as for trees, $N/M$, where $N$ is the number of correctly classified training cases and $M$ the number of incorrectly classified ones by a rule). The lift 2.0 is the result of dividing the rule's estimated accuracy by the relative frequency of the predicted class in the training set.

The class assignment, -> class 0 [0.978], contains a value between 0 and 1 that indicates the confidence with which this prediction is made. If a case is not covered by any rule, then it is used as a so-called default class which, in this example, is Default class: 1 (1 means *positive*; can be found in the output).

After the list of rules, the output contains an evaluation of the rule set accuracy as well as a list of relevant attributes, as is done for trees. For example:

```
Evaluation on training data (900 cases):

            Rules
        ----------------
      No       Errors

      20   121(13.4%)   <<


     (a)    (b)     <-classified as
     ----   ----
     358     85     (a): class 0
      36    421     (b): class 1


     Attribute usage:

   100.00% waste
    84.00% life
    30.67% tale
    25.89% pages
    24.78% not
    22.56% dont
    21.44% wonderful
          . . .
```

To obtain the result of testing the generated rules with 100 external samples, the same procedure as followed for the tree needs to be carried out. It is necessary to get class labels using the created model and test data:

```
predictions <- predict(c5_rules, test_data, type="class")
```

The rest of the calculations are the same as in the case of using a decision tree. The results of testing are shown here:

```
[1] "Confusion matrix"
> print(cm)
```

```
          predictions
test_labels  0   1
          0 32  25
          1 11  32
```

```
[1] "Detailed classification metrics"
   precision    recall    f1
0 0.7441860 0.5614035 0.64
1 0.5614035 0.7441860 0.64
[1] "Accuracy: 0.64"
[1] "Macro-averaged metrics"
precision    recall          f1
0.6527948 0.6527948 0.6400000
```

```
[1] "Micro-averaged metrics"
precision    recall          f1
0.6655896 0.6400000 0.6400000
```

Again, it is obvious that testing samples give a higher accuracy error, 36%, while the training samples are better with an error of only 13,4%. However, the discussion about the tree is valid also for the rules – trees and rules are just two different representations of knowledge revealed from data.

The not quite convincing classification results in the demonstration presented above do not mean that decision tree or rules are inferior algorithms. There are a huge number of successful applications based on trees and rules, including text mining. The practice shows that varied data may need different algorithms and no algorithm is universal [282] – what works for certain datasets does not have to be convenient for others and vice versa. Sometimes, a user has to select a less perfect algorithm just because it has significantly lower computational complexity and can be used in real time. On that account, machine learning is an open field for experimenting.

# Chapter 8

# Random Forest

## 8.1  Introduction

*Random forest*, as the name suggests, is a classification method based on applying a group of decision trees. It belongs to classification methods which use simultaneous voting by more than one trained expert algorithm and the result is given by their majority (for regression tasks, by their average).

The random forest idea is based on an attempt to suppress over-fitting of (as well as correlation between) the trees in a group (forest). This suppression is achieved by randomly selecting attributes for each (sub)tree at each splitting node in addition to randomly selecting subsets of training samples using *bagging* (*bootstrap aggregation*); also refer to [32, 72].

### 8.1.1  Bootstrap

*Bootstrap* is a statistical method based on random selection with replacement. This means that a 'set' created by this selection of samples may contain some examples multiple times (so, strictly speaking, it cannot be called a set).

The idea is that when some samples are randomly selected multiple times, some others are never selected – and those unselected samples may therefore be used as a test set.

The basic procedure is as follows. The data having $n$ samples are subjected to $n$ selections with replacement. The chance that an example will not be selected in any of $n$ attempts is based on that the probability of being chosen is $1/n$ and – inversely – of not being chosen, $1 - 1/n$.

If the selection is repeated $n$ times, then multiplication of the probabilities provides the estimate of how many times no selection will be made:

$$(1 - 1/n)^n = e^{-1} \doteq 0.3678794412\ldots(\text{for } n \to \infty) \tag{8.1}$$

For example, for 1000 selections with replacement, it is $(1 - 1/1000)^{1000} = 0.3676954248\ldots \doteq 36.77\%$.

Thus, there are two sets of samples now: 63.23% for training and 36.77% for testing, or, in other words, approximately 2/3 for training and 1/3 for testing. Sometimes this method is called '0.632 bootstrap.' Note that in this case, neither the training nor the testing set eventually contains multiplied samples.

The basic essence of classification by a forest of individually trained decision-making trees is shown in Figure 8.1. Defining the number of trees in a forest, $n$, and applying the *bagging* method to the original set of available training samples, $D$, a testing subset $d_0$ as well as a group of $n$ training subsets $d_i$, $1 \le i \le n$, are generated, where $d_i \subset D$, $D = \cup_{i=0}^n d_i$.

Each of the $n$ trees is then trained using its own $d_i$ and works as a classifier. Because each of the trees in the created forest is usually more-or-less different

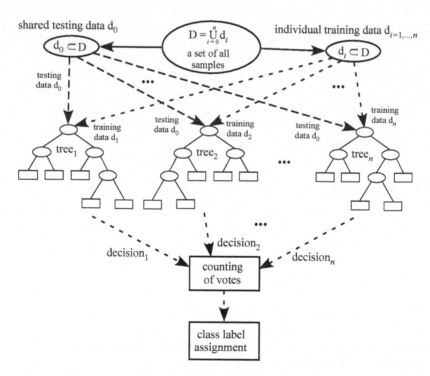

**Figure 8.1:** Random trees – the main idea of a classifier group composed of $n$ decision trees that are trained by randomly selected subsets of training samples.

from the rest, it is quite possible that their classification of a given unlabeled item differs as well – the reason is the fact that those trees are based on more-or-less different subsets of attributes. As a result, the relevance of the same attributes may be quite different for each tree. The resulting assignment of the class label to the classified item is finally determined by the prevailing majority in the vote.

And the final question: What is the best number *n* of trees? The optimal number is achieved when adding more trees does not improve the classification. This value can be found experimentally for practical applications – direct determination is generally not supported by theory. Two requirements should be met: the lowest classification error and the smallest number of trees.

## 8.1.2 Stability and Robustness

In order to avoid '*too good*' or '*too bad*' a random selection (which is also quite possible), it is advisable to repeat experiments, each time with a different seed of the (pseudo)random number generator (better still, is to use a real random number source, which may not always be achievable).

In general, each testing can provide a more or less different estimated future classification error, so the expected error value is calculated as the average (it also applies to cross-validation). The variance of the average expected error provides information on the stability of the algorithm with respect to data composition. To achieve good stability (algorithm robustness), the smallest possible variance is ideal.

This mainly affects the data itself, but different algorithms can give different results from the same point of view, though the average error may be almost the same. The test results are therefore another aspect of the algorithm selection and robustness, which is a fundamental truth for decision trees that can very easily suffer from instability due to even small changes in training examples.

## 8.1.3 Which Tree Algorithm?

In principle, what kind of decision tree is used is not decisive. Often, implementations and software packages with *CART* (*Classification And Regression Tree*; for details see [36]) are used, but *c5* or others could be employed as well. For practical reasons and simplicity, the *CART* system is used in the following demonstration because it is easily available in R.

## 8.2 Random Forest in R

This chapter presents a brief demonstration of applying the random forest method in R to the same sample of data as in the previous cases (the reviews of books). The demonstrated use of the random forest method here, based on a popular R-package *randomForest*, comes from the original implementation in the Fortran programming language by statistician Leo Breiman (January 27, 1928 - July 5, 2005). For more details, see [33, 34, 35]. The detailed description of *randomForest* parameters in R can be found in [171].

```r
library(randomForest)

# loading the data
data <- read.csv("BookReviews.csv", header=TRUE)

# converting the data to a matrix
dtm <- as.matrix(data)

# removing the object data (unnecessary from now)
rm(data)

# number of instances
n <- dim(dtm)[1]

# number of attributes
m <- dim(dtm)[2]

# indexes of 100 testing samples
number_of_test_samples <- 100
test_index <- 1:number_of_test_samples

# selecting training samples
train_data <- dtm[-test_index, 1:(m-1)]
train_labels <- dtm[-test_index, m]

# selecting testing samples
test_data <- dtm[test_index, 1:(m-1)]
test_labels <- dtm[test_index, m]

# training a random forest model
rf_model <- randomForest(train_data, as.factor(train_labels))
predictions <- predict(rf_model, test_data, type="response")

# creating a confusion matrix
cm <- table(test_labels, predictions)
print("Confusion matrix")
print(cm)
```

```r
# number of instances
n <- sum(cm)

# number of correctly classified instances for each class
correct <- diag(cm)

# numbers of instances in each class
instances_in_classes <- apply(cm, 1, sum)

# numbers of each class predictions
class_predictions <- apply(cm, 2, sum)

# accuracy
accuracy <- sum(correct)/n

# precision per class
precision <- correct/class_predictions

# recall per class
recall <- correct/instances_in_classes

# F1-measure per class
f1 <- 2 * precision * recall / (precision + recall)

# printing summary information for all classes
df <- data.frame(precision, recall, f1)
print("Detailed classification metrics")
print(df)
print(paste("Accuracy:", accuracy))

# macroaveraging
print("Macro-averaged metrics")
print(colMeans(df))

# microaveraging
print("Micro-averaged metrics")
print(apply(df, 2, function (x)
                weighted.mean(x, w=instances_in_classes)))

# plotting the error ratos or MSE of the randomForest object
plot(rf_model, col=c(1,2,4), lwd=1.5, lty=1)
rf_legend <- if (is.null(rf_model$test$err.rate)) {
                colnames(rf_model$err.rate)
```

```
            } else {
                colnames(rf_model$test$err.rate)
            }
legend("topright", cex=1, legend=rf_legend, lty=1,
        col=c(1,2,4), horiz=TRUE)
```

As a result, the script returns the following output:

```
[1] "Confusion matrix"
            predictions
test_labels  0  1
           0 38 10
           1  9 43


[1] "Detailed classification metrics"
   precision    recall         f1
0 0.8085106 0.7916667 0.8000000
1 0.8113208 0.8269231 0.8190476
[1] "Accuracy: 0.81"

[1] "Macro-averaged metrics"
precision    recall         f1
0.8099157 0.8092949 0.8095238

[1] "Micro-averaged metrics"
precision    recall         f1
0.8099719 0.8100000 0.8099048
```

The graph in Figure 8.2 shows the classification accuracy error dependence on the number of trees in the random forest. The graph contains curves for the positive class (marked as 1), negative class (marked as 0), and the OOB (Out-Of-Bag, the standard mean error for both classes measured using the test samples).

As a reader can see, the classification accuracy is better than 80%. For a small number of random trees, the classification error is relatively high, especially at the beginning: for just one random tree, the error is nearly 50%, which corresponds to a random guess. With the increase in forest size, the error decreases, very quickly initially. After reaching a certain number of trees, around the vicinity of 300, the classification no longer improves. Thus, 300 random trees is perhaps the optimal solution.

Such a number of trees might look very high – however, each tree is generated by a limited number of attributes (words) because it uses only a relatively small subset of all training samples, therefore a limited vocabulary. As a result, each tree is also relatively small, perhaps not a very 'smart' expert but nonetheless, a real expert in its narrow area.

rf_model

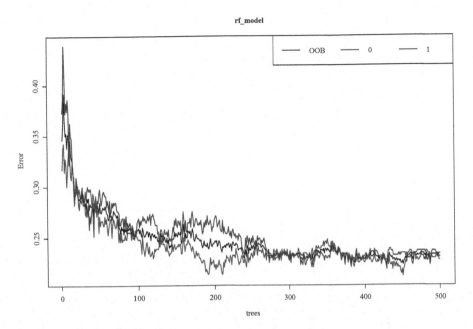

**Figure 8.2:** Classification accuracy dependence on the number of trees (around 200 trees, OOB is in the middle, 0 bottom, 1 top).

However, very good results can be achieved by gathering limited experts whose areas somehow overlap with each other, which is one of the main and successful ideas of the method used and presented here. In recent times, the popularity of random forests in many practical application areas has shown an increase.

# Chapter 9

# Adaboost

## 9.1 Introduction

*Adaboost* (*Adaptive Boosting*) [94, 95, 112] belongs to a group of methods, or rather meta-algorithms, that do not focus on optimizing one particular algorithm – instead, *Adaboost* uses an ensemble of algorithms, which individually do not provide excellent results but, together, give very good results. The main idea is similar to the above mentioned *Random Forests* algorithm: Simple, slightly above average experts can, together, provide an excellent result by voting and thus overcoming the success of any single smart expert. *Adaboost* is based on weighting training samples as described lower.

## 9.2 Boosting Principle

In its ensemble, a boosting meta-algorithm usually employs the same type of an algorithm called *weak learner*. The weak learner is an algorithm for which it is quite sufficient that (considering two classes) its classification error is arbitrarily less than 50% (however, this is only a minimal request).

To improve classification results, additional classifiers with the same property are added gradually to upgrade the overall outcome. Each such classifier may work well on only 'its own' randomly selected part of the training samples, while leaving other samples to colleagues – the ensemble's classification ability is thus enhanced.

The starting requirement is the determination of the number of weak learners, $n_{wl}$ which is related to the random division (by *bagging*, for example) of the training set into the appropriate number of training subsets, $n_{wl}$ one for each learner.

There is no unambiguous algorithm to determine the value of $n_{wl}$ in advance. The general requirement is that every sample from the training set $D$ is involved in the training process and further, the number of training samples for each learner would be approximately the same, given that every learner has an equal vote in the final voting. Each trained learner then provides its classification of an unlabeled item and the result is given by majority.

The principle is quite simple, but it has, unfortunately, some difficulties in practice. Easier classifying tasks are typically characterized by the fact that the first trained classifier just works correctly on the vast majority of samples $m_1$, where $m_1 >> m_i$, $i = 2....,n_{wl}$. The assumption that only a small number of samples remains to be covered by the following necessary classifiers to complete the task results in insufficient training due to the lack of training samples. As a consequence, not all training samples available in $D$ are used and the result is not optimal, given the circumstances.

The applied solution is usually to repeat the whole boosting procedure a few times to ensure the best value for $m_1$, approximately equal distribution of number of samples in training subsets, and, if possible, utilization of all training samples. In order to do this, various heuristics are used to try to keep up requirements.

## 9.3 Adaboost Principle

Nowadays, *Adaptive Boosting* is probably the most widely used boosting option. It allows the addition of weak learners until a certain desired low classification error value is reached. Each training sample is given a weight which determines the probability of its selection from $D$ into the training subset of each individual classifier from the created ensemble. If a training sample is classified correctly, its chances of re-selection by the following classifier decrease; otherwise it grows. Thus, *Adaboost* focuses on difficult samples.

Initially, the samples receive the same weight value. In every iteration step $k$, $1 \leq k \leq k_{max}$, a particular training set is generated by the sample selection based on the weights which the probability of being selected. Each classifier, $C_k$, is trained on its $k$'s set.

In the next step, with the help of $C_k$, the weights of incorrectly classified samples are increased and weights of samples classified correctly are reduced.

In line with their weights and based on the new selection probability distribution, the remaining samples are then used to create another classifier $C_{k+1}$, and the process continues iteratively until the predetermined maximum number of weak learners, $k_{max}$, is reached, no next training samples are available, or the lowest desired classification error of the ensemble is reached.

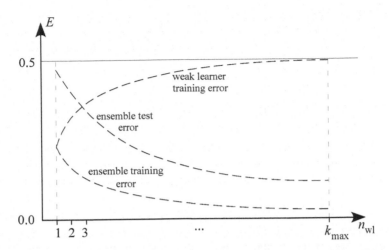

**Figure 9.1:** The classification error of a classifying ensemble depending on the number of weak learners.

Figure 9.1 illustrates the typical progress of the training process with the gradual addition of more and more weak learners (the axis $n_{wl}$). The first weak learner shows the highest classification error $E$ (recall that it must be always less than 50% for two classes) on both training and test data. After adding another weak learner (that focuses on samples not mastered by previous weak learners), the overall error of the ensemble always drops somewhat on both the training and test data, although the error rate of the gradually added individual weak learners typically progressively increases due to their focus on increasingly difficult examples. As long as every weak learner works with an error value less than 50%, the weighted decision by the entire voting ensemble guarantees a drop in error on the *training* data, and commonly on the testing data as well.

In theory, the presented curves represent an exponential decrease and increase in error values.

In any case, it is necessary to be cautious because increasing the number of classifiers $k_{max}$ in an ensemble can lead to undesirable over-training (loss of ability to generalize due to overfocusing of classifiers to recognize only "their" specific training data). However, many simulation experiments have shown that this rarely happens, even for extremely high values of $k_{max}$.

A brief and well-explained issue of the *Adaboost* method can be found, for example, in [76]. To summarize, the following empirical basic rule applies to the practical application of *Adaboost*:

Boosting improves classification only if individual classifiers in an ensemble provide better than just random results – unfortunately, this cannot be guaranteed in advance.

Despite the absence of *a priori* guarantee, the method provides very good results in many real-world problems, which is why it is so popular today and used frequently, represented by various implementations, including R.

## 9.4   Weak Learners

What classification algorithm should be used as a *weak learner*? Given the basic requirement for a weak learner to achieve better than completely random results, choosing the right algorithm is not critical. Most implementations, including the one demonstrated here in R, use decision trees, especially their simplest variant known as *stump*.

The stump algorithm, as its name indicates, is a tree limited to the root node from which the branches lead directly to the leaves. The root contains a test for the value of one of the attributes, preferably one that reduces as much as possible the heterogeneity of a set containing a mixture of examples from different classes. Hence, a better than random selection results in more homogeneous subsets.

Note that it is not excluded that some stumps may share the same attribute. In such cases, the root can test the shared attribute using the same or quite different values, depending on the specific situation.

Figure 9.2 shows an example of the stump-tree. For a given task, one of the available attributes, $x_i$, is chosen to decide which label should be assigned to a classified item, based on the specific value of that $x_i$.

Because the attribute classification relevance may change during the gradual addition of weak learners (as well as the removal of examples from the original training set), different stumps arise in terms of the tested attributes in the roots. Hence, it may also be an advantage that the list of such attributes is a secondary (or eventually might be even primary) result of the classification, thus contributing to the knowledge extracted from the analyzed data.

The unquestionable advantage of using stumps is that due to their simplicity they do not need long and complicated training. Given the principles of the *Adaboost* method, the desired result can be achieved simply by using a sufficient amount of stumps, which naturally depends on a sufficient number of quality training samples covering the required classes.

Figure 9.3 illustrates the *Adaboost*-based process of classification. Gradually, $k_{max}$ stumps were trained using $k_{max}$ mutually disjunctive subsets of the original set of training samples.

Having two classes, each stump uses one of the attributes for its decision on whether the unlabeled item belongs to class $C_1$ or $C_2$. The final decision is given by voting majority.

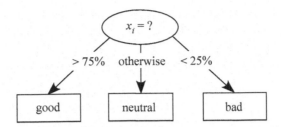

**Figure 9.2:** An example of *stump*. The root tests the attribute $x_i$ and, according to the result, assigns one of three possible classes to an unlabeled item.

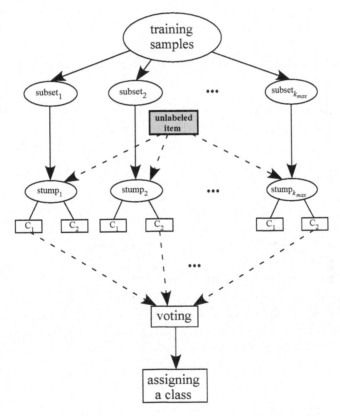

**Figure 9.3:** The illustration of training and classification by *Adaboost*.

## 9.5    Adaboost in R

The demonstrated application of *Adaboost* uses the library called ada, which must first be installed. The results of two different tree-based weak learners are shown for the purpose of illustration: a *decision tree* (the `rpart` library, Recursive Partitioning and Regression Trees), and *stump* (in the R-code below, see the variable *maxdepth* = 1 for stump).

The R code for running the described *Adaboost* meta-algorithm:

```
library(ada)
library(rpart)

# loading the data
data <- read.csv('BookReviews.csv', header=TRUE)

# converting the data to a matrix
dtm <- as.matrix(data)

# removing the object data (unnecessary from now)
rm(data)

# number of instances
n <- dim(dtm)[1]

# number of attributes
m <- dim(dtm)[2]

# indexes of 100 testing samples
number_of_test_samples <- 100
test_index <- sample(n, number_of_test_samples)

# selecting training samples
train_data <- dtm[-test_index, 1:(m-1)]
train_labels <- dtm[-test_index, m]

# selecting testing samples
test_data <- dtm[test_index, 1:(m-1)]
test_labels <- dtm[test_index, m]

result <- NULL

for (it in c(1, 5, seq(10, 100, 10))) {
    # with trees
    ada_model <- ada(as.data.frame(train_data),
                     train_labels,
                     type="real",
                     iter=it,
```

```
                    nu=0.05,
                    bag.frac=1)
predictions_ada <- predict(ada_model,
                           as.data.frame(test_data),
                           type="vector")
# with stumps
ada_model_st <- ada(as.data.frame(train_data),
                    train_labels,
                    type="real",
                    iter=it,
                    nu=0.05,
                    bag.frac=1,
                    # rpart.control for stumps
                    control = rpart.control(maxdepth=1,
                           cp=-1, minsplit=0, xval=0))
predictions_ada_st <- predict(ada_model_st,
                              as.data.frame(test_data),
                              type="vector")


# creating confusion matrices
cm_ada <- table(test_labels, predictions_ada)
cm_ada_st <- table(test_labels, predictions_ada_st)


# number of instances
n <- sum(cm_ada)


# number of correctly classified instances for each class
correct_ada <- diag(cm_ada)
correct_ada_st <- diag(cm_ada_st)


# numbers of each class predictions
class_predictions_ada <- apply(cm_ada, 2, sum)
class_predictions_ada_st <- apply(cm_ada_st, 2, sum)


# accuracy
accuracy_ada <- sum(correct_ada)/n
accuracy_ada_st <- sum(correct_ada_st)/n


# precision per class
precision_ada <- correct_ada/class_predictions_ada
precision_ada <- round(precision_ada, 4)


precision_ada_st <- correct_ada_st/class_predictions_ada_st
```

```
precision_ada_st <- round(precision_ada_st, 4)

# storing the results of current iteration
result <- rbind(result,
                data.frame(it, accuracy_ada, accuracy_ada_st,
                           precision_ada["1"],
                           precision_ada["0"],
                           precision_ada_st["1"],
                           precision_ada_st["0"]))
}
# printing classification performance measures
# for individual iterations
rownames(result) <- NULL
colnames(result) <- c("It.",
                       "Acc1", "Acc2",
                       "Prec1(pos)", "Prec1(neg)",
                       "Prec2(pos)", "Prec2(neg)")
print(result)

# plotting the development of classification errors
plot(ada_model)
plot(ada_model_st)

# printing the model information and final confusion
# matrix for submitted training data
print("Adaboost with trees ")
print(ada_model)
print('Adaboost with stumps ')
print(ada_model_st)
```

The values of classification performance metrics for the adaboost algorithm with trees and stumps will appear on the output (Acc1, Prec1 etc. are the values for the algorithm with trees, the names ending with 2 are related to the algorithm with stumps):

|   | It. | Acc1 | Acc2 | Prec1(pos) | Prec1(neg) | Prec2(pos) | Prec2(neg) |
|---|-----|------|------|------------|------------|------------|------------|
| 1 | 1   | 0.63 | 0.49 | 0.6731     | 0.5833     | 0.5370     | 0.4348     |
| 2 | 5   | 0.63 | 0.52 | 0.6667     | 0.5870     | 0.5686     | 0.4694     |
| 3 | 10  | 0.68 | 0.58 | 0.7170     | 0.6383     | 0.6512     | 0.5263     |
| 4 | 20  | 0.67 | 0.61 | 0.7115     | 0.6250     | 0.6600     | 0.5600     |
| 5 | 30  | 0.68 | 0.63 | 0.7347     | 0.6275     | 0.6731     | 0.5833     |
| 6 | 40  | 0.72 | 0.66 | 0.7755     | 0.6667     | 0.6909     | 0.6222     |

| | | | | | | | |
|---|---|---|---|---|---|---|---|
| 7 | 50 | 0.70 | 0.69 | 0.7551 | 0.6471 | 0.7069 | 0.6667 |
| 8 | 60 | 0.76 | 0.69 | 0.7925 | 0.7234 | 0.7000 | 0.6750 |
| 9 | 70 | 0.74 | 0.67 | 0.7843 | 0.6939 | 0.6833 | 0.6500 |
| 10 | 80 | 0.73 | 0.68 | 0.7800 | 0.6800 | 0.6885 | 0.6667 |
| 11 | 90 | 0.74 | 0.70 | 0.7843 | 0.6939 | 0.6984 | 0.7027 |
| 12 | 100 | 0.75 | 0.69 | 0.7885 | 0.7083 | 0.6935 | 0.6842 |

An interested reader can also look at additional output information which, while not presented here, can be seen on the RStudio console after the end of computation. In addition, two graphs (see RStudio, Plots, in the lower right quarter of console) created by the commands plot(ada_model) for the tree and plot(ada_model_st) for the stump are available. The graphs in Figure 9.4 and Figure 9.5 demonstrate the course of classification errors for the training samples during iterations.

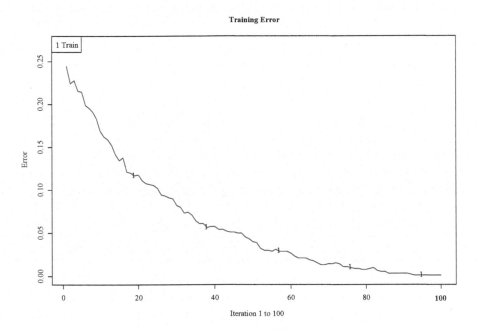

**Figure 9.4:** *Adaboost with trees*: the iterative accuracy error reduction.

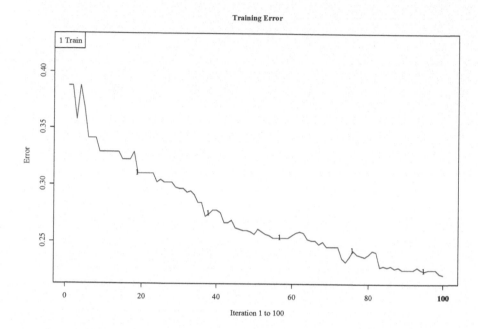

**Figure 9.5:** *Adaboost with stumps*: the iterative accuracy error reduction.

It is evident that the full tree works better than the simpler stump; however, in the end, the resulting classification accuracy for the tree variant (75%) is only 6% better than that for the stump (69%). As can be seen here, the stump suffers from poor classification of negative samples, which should perhaps be refined in order to match the positive samples from the information point of view.

# Chapter 10

# Support Vector Machines

## 10.1   Introduction

*Support Vector Machines, SVM,* is another successful algorithm that has proven and demonstrated excellent results in many different real-world applications. Its foundations date back to relatively early times, beginning in 1963, when the Russian (Soviet) mathematician Vladimir Vapnik (along with Alexey Chervonenkis, Alexander Lerner, and others) published works on statistical learning and pattern recognition.

These works gradually gave rise to one of the current successful algorithms based on the theory of linear algebra [31]. In principle, *SVM* is a classification algorithm aimed at separating two classes using a hyperplane – a linear classifier. There are several ways to place a hyperplane (or a straight line in two-dimensional space) that separates elements of two classes. The problem is that if such an option exists, one can generally create infinitely many such boundaries for a given set of training examples; see Figure 10.1.

The figure shows here only three of the infinitely many possible separating linear boundaries, *a, b,* and *c,* found by some method. For the training data (which does not include that misclassified item), all three boundaries are perfect, so any of them can be selected. However, if a new element belonging to the white ring class appears, two of the initially set boundaries (*a* and *b*) put it wrongly in the black rings. So, is the boundary marked *c* optimal? Or not? The question is whether an *optimal* solution can be found based on training examples to minimize such potential future misclassifications as much as possible – hence the *optimum hyperplane* is such that the minimum value of the distance of the points from that plane is as large as possible.

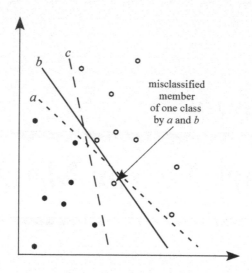

**Figure 10.1:** Some of possible linear boundaries can wrongly separate a class member.

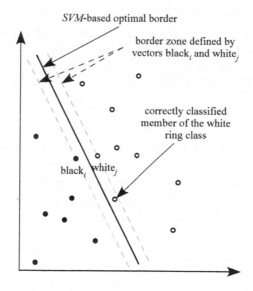

**Figure 10.2:** The optimal border found by *SVM* passes through the center of the border zone defined by the $black_i$ and $white_j$ support vectors.

*Support Vector Machines* algorithm can be offered as the right answer – see the illustration Figure 10.2, which uses the same *black* and *white* samples as Figure 10.1.

First, according to the *SVM*'s idea, it is necessary to find the widest zone (belt) separating both classes. Suitably selected examples called *support vectors* determine the edges of this zone.

Next the optimum boundary (hyper)plane can be placed through the center of the zone. Of course, without knowing all the possible cases (which can theoretically be infinite), this may not guarantee absolute faultlessness for the future classifications of the cases that are not available for the training process, but it increases the chance that even very near-border cases will be classified correctly.

## 10.2    Support Vector Machines Principles

The linear function behaviour, unlike the nonlinear one, has its shape firmly predetermined. This simplifies the whole situation in the application of linear separation because the nonlinear pattern can be arbitrary and its correct selection for a boundary may not often be quite obvious. Therefore, if possible (and also in accordance with Occam's razor), it is appropriate to use the linear function.

The principles of the *Support Vector Machines* algorithm are based on a relatively large non-trivial mathematical apparatus combining a number of sectors (especially linear algebra, kernel functions, nonlinear optimization). Unfortunately, due to limitations of space, it is not possible to reproduce details here. The reader may try the example presented below to apply *SVM* to the sample data, play with parameters (for example, the type of kernel function – linear, polynomial, radial) and, in the case of deeper interest in the basics, get familiar with the details in the literature – some links to excellent resources are presented in References.

### 10.2.1    Finding Optimal Separation Hyperplane

The first problem, as indicated with the help of Figure 10.1 and Figure 10.2, is the problem of locating the separating hyperplane as best as possible by finding the boundary zone – that is, finding, among all the samples, those whose combinations would create the widest zone. Obviously, this is computationally very demanding and it is not possible to do it in practice by trying all possible combinations because of computational complexity – for some applications, there might be millions of training samples, or more.

Thus, certain simplifying techniques and heuristics must be used to overcome this high complexity. Various implementations of *SVM* use the optimization method known as *quadratic programming* (optimizing a quadratic function of variables with linear constraints); for an example, see [24], which is a special case of *nonlinear programming*.

## 10.2.2 Nonlinear Classification and Kernel Functions

The second problem is that linear boundaries are often not sufficient because the task inherently requires a nonlinear boundary – and this leads to difficulty in determining a shape of the nonlinear boundary, being possibly stuck in local extremes as it is known for some alternative algorithms (typically, artificial neural networks with neurons having a nonlinear transfer function). A surprising solution may seem to be that when a curve cannot be straightened in a given space to provide a linear separation, it is possible to adjust, "incurvate," the task space somehow in order to use a hyperplane.

This problem relating to nonlinear classification is solved by so called *kernel transformation* based on *kernel functions* [30, 243], which is the *SVM*'s key to adapting the properties of the space under investigation: the transformation of the original space of the data attributes into the space of transformed attributes – typically, the space of higher dimension. This kernel transformation makes it possible to convert the initially linearly non-separable task into a linearly separable one, to which an optimization algorithm can be applied in order to find the separating hyperplane.

The transformation process is illustrated in Figure 10.3 (from one-dimensional to two-dimensional space) and Figure 10.4 (from two-dimensional to three-dimensional space).

In one-dimensional space, there is evidently no possibility of separating black and white circle classes from one another using a hyperplane (here, a point somewhere on the $x_1$ axis). Only after the transformation into a two-dimensional space, where a new artificial coordinate $x_2$ is created as a function of $x_1$, $x_2 = f(x_1)$, and the points are distributed on a parabola (kernel function), can the necessary boundary zone be found and in the middle of it, a separating boundary (the grey belt) of both classes.

Similarly, it is for the transformation from two-dimensional space (one class is bounded by an oval) into three-dimensional space with a new artificial dimension $x_3 = f(x_1, x_2)$, where the white and black circles are placed on the surface of the paraboloid (kernel function).

The transformation of vectors (training samples) from the Euclidean space, where it is not possible to separate both classes linearly, into another Euclidean space, where it is already possible, is done by kernel functions. The essence is that it is not necessary to know the representation of the $\Phi$ transformation function, but it is sufficient to calculate only the scalar product of the $\Phi$ function applied to the transformed vectors, using the kernel function $K$:

$$K(\vec{x}_i, \vec{x}_j) = \Phi(\vec{x}_i) \cdot \Phi(\vec{x}_j).$$

For example, for the linear transformation, $K(\vec{x}_i, \vec{x}_j) = \vec{x}_i \cdot \vec{x}_j$, for the polynomial function of degree $d$ it is $K(\vec{x}_i, \vec{x}_j) = (\vec{x}_i \cdot \vec{x}_j + 1)^d$, and so like.

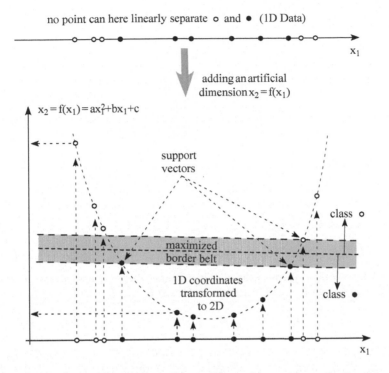

**Figure 10.3:** The Support Vector Machine (SVM) principle of creating the linear boundary between classes using the specific SVM transformation into a space with a larger number of dimensions. The above example illustrates transformation from 1D to 2D space.

Indeed, the use of a scalar product greatly simplifies and accelerates the computational process. A reader can find additional information in [59, 111].

## 10.2.3 Multiclass SVM Classification

The third problem might be that *SVM* procedure can separate only two classes (one above, the other, below a hyperplane) from each other. However, in reality, there may possibly be more than two classes.

At first glance, there are two simple straightforward methods based on the transformation to the binary case: *one-against-one* (all individual couples are trained against each other) and *one-against-all* (a particular class is trained against the rest of classes).

The former method suffers mainly from high computational complexity as well as from scale variability of the confidence values that may differ between the binary classifiers, while the latter method typically suffers from the unbalanced number of training samples even if the individual training sets are balanced.

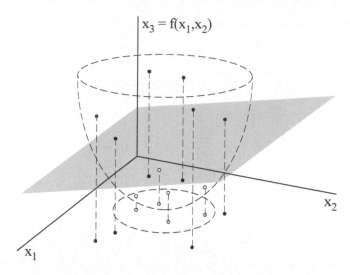

**Figure 10.4:** Creating a hyperplane (marked here in gray) that separates two classes using the transformation from the original 2D to artificial 3D space.

Apart from the two above-mentioned methods, extensions to handle the multiclass classification case have been proposed. In these extensions, additional parameters and constraints are added to the optimization problem to handle the separation of the different classes. Details and additional references may be found in literature, for example [274].

## 10.2.4   SVM Summary

Modern *SVM*-based classification systems can solve all the mentioned issues. Even if *SVM* is generally computationally demanding, its result is usually worthwhile.

One of *SVM*'s big advantages (unlike, for example, the $k$-NN algorithm mentioned earlier) is that after training, only the found support vectors need to be stored for future classifications, which in most applications is typically just a fraction of all provided training samples. The real-time classification then proceeds very fast.

Because of the linearity, training always finds a global (optimal) solution with reproducible results, independent of the choice of a starting point (unlike, for example, artificial neural networks).

On the other hand, there are also some problems and limitations. The correct function of the *SVM* depends largely on the appropriate choice of kernel function, which is not always easy to find. For very large numbers of support vectors, there is a problem with speed and memory requirements. Appropriate algorithms for their solution, namely because of the current group of problems connected with *Big Data*, are still under investigation.

## 10.3  SVM in R

Using the same set of training samples, the classification with one of the available implementations of *SVM* in the R-system is demonstrated here. There are several packages/libraries available. An implementation from the package e1071 was selected for the demonstration of the *SVN* functionality because it provided the best classification results for the demonstration data. The applied script follows below. After the preparation of training and test data, the model based on training samples is created using the function svm_model(). The linear kernel function selected here; however, an interested reader may try alternative kernel functions (radial basis function RBF, sigmoidal, or polynomial). The parameter type="C-classification" is used for binary classifications, which is also the presented example. The complete parameter description of e1071 is available in [193].

The advanced implementation provides rich capabilities, but it takes some time to study and test combinations of different parameters depending on the data being analyzed – for reasons of space, it is not possible to provide detailed instructions here, but the Internet can help find the necessary information.

```
library(e1071)

# loading the data
data <- read.csv("BookReviews.csv", header=TRUE)

# converting the data to a matrix
dtm <- as.matrix(data)

# removing the object data (unnecessary from now)
rm(data)

# number of instances
n <- dim(dtm)[1]

# number of attributes
m <- dim(dtm)[2]

# indexes of 100 testing samples
```

```
number_of_test_samples <- 100
test_index <- 1:number_of_test_samples

# selecting training samples
train_data <- dtm[-test_index, 1:(m-1)]
train_labels <- dtm[-test_index, m]

# selecting testing samples
test_data <- dtm[test_index, 1:(m-1)]
test_labels <- dtm[test_index, m]

# training a SVM model for classification
svm_model <- svm(train_data,
                 train_labels,
                 type="C-classification",
                 kernel="linear")

# calculating class predictions
predictions<-predict(svm_model, test_data, type="class")

# creating a confusion matrix
cm <- table(test_labels, predictions)
print("Confusion matrix")
print(cm)

# number of instances
n <- sum(cm)

# number of correctly classified instances for each class
correct <- diag(cm)

# numbers of instances in each class
instances_in_classes <- apply(cm, 1, sum)

# numbers of each class predictions
class_predictions <- apply(cm, 2, sum)

# accuracy
accuracy <- sum(correct)/n

# precision per class
precision <- correct/class_predictions
```

```
# recall per class
recall <- correct/instances_in_classes

# F1-measure per class
f1 <- 2 * precision * recall / (precision + recall)

# printing summary information for all classes
df <- data.frame(precision, recall, f1)
print("Detailed classification metrics")
print(df)
print(paste("Accuracy:", accuracy))

# macroaveraging
print("Macro-averaged metrics")
print(colMeans(df))

# microaveraging
print("Micro-averaged metrics")
print(apply(df, 2, function (x)
                weighted.mean(x, w=instances_in_classes)))
```

Output:

```
[1] "Confusion matrix"
          predictions
test_labels  0  1
          0 40 17
          1 12 31

[1] "Detailed classification metrics"
   precision    recall        f1
0 0.7692308 0.7017544 0.7339450
1 0.6458333 0.7209302 0.6813187
[1] "Accuracy: 0.71"

[1] "Macro-averaged metrics"
precision    recall        f1
0.7075321 0.7113423 0.7076318

[1] "Micro-averaged metrics"
precision    recall        f1
0.7161699 0.7100000 0.7113157
```

The following demonstration of SVM uses cross-validation for testing. The results of individual steps show similar results and the standard error of mean, which is 1.49, indicates that SVM is a relatively robust algorithm for the given data robust algorithm.

```
Cross-validation fold: 1
 Accuracy: 0.71
 Precision(positive): 0.6458333
 Precision(negative): 0.7692308
Cross-validation fold: 2
 Accuracy: 0.75
 Precision(positive): 0.8
 Precision(negative): 0.7090909
Cross-validation fold: 3
 Accuracy: 0.73
 Precision(positive): 0.76
 Precision(negative): 0.7
Cross-validation fold: 4
 Accuracy: 0.78
 Precision(positive): 0.74
 Precision(negative): 0.82
Cross-validation fold: 5
 Accuracy: 0.65
 Precision(positive): 0.6666667
 Precision(negative): 0.6346154
Cross-validation fold: 6
 Accuracy: 0.78
 Precision(positive): 0.7924528
 Precision(negative): 0.7659574
Cross-validation fold: 7
 Accuracy: 0.69
 Precision(positive): 0.6136364
 Precision(negative): 0.75
Cross-validation fold: 8
 Accuracy: 0.66
 Precision(positive): 0.6792453
 Precision(negative): 0.6382979
Cross-validation fold: 9
 Accuracy: 0.73
 Precision(positive): 0.7608696
 Precision(negative): 0.7037037
```

```
Cross-validation fold: 10
 Accuracy: 0.77
 Precision(positive): 0.7931034
 Precision(negative): 0.7380952
Average Accuracy: 72.5%
SEM: 1.49
```

# Chapter 11

# Deep Learning

## 11.1   Introduction

This final chapter of the basic current methods of classification in the text mining sector focuses very briefly on one of the newest directions called *Deep Learning*; see for an example, an extensive detailed book [103]. This direction is inspired by the broad possibilities of artificial neural networks, which – as one might say – rank among the oldest and truly traditional algorithms.

With the rapid development of computer technology performance (speed, memory), one of the original disadvantages of neural networks – their computational demands from the hardware point of view – is gradually being overcome, opening up their further availability to wider utilization in other challenging real-world problem areas.

Which tasks are directly destined for deep learning? They typically include learning high-level attributes, or transformation of attributes, especially under the conditions when the attributes cannot be extracted or defined manually, the dimensionality of data is very high, or classical learning algorithms do not seem to work well enough or at all. Some of these conditions, if not all of them, particularly the high dimensionality apparently apply to text mining too, as the previous chapters might have suggested.

It is the characteristic of deep learning that it selects or creates relevant symptoms which may not at all be obvious to the problem being solved. The following illustration in Figure 11.1 simply illustrates the principle difference between traditional machine-learning algorithms and deep learning.

The deep learning procedure currently includes mainly multi-layer feed-forward networks, deep convolutional neural networks, deep-belief networks, and recurrent neural networks for learning from sequential data.

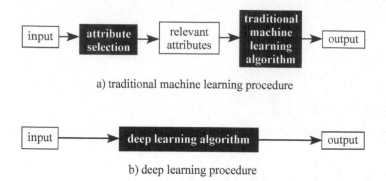

a) traditional machine learning procedure

b) deep learning procedure

**Figure 11.1:** Unlike the *traditional* machine-learning approach, *deep learning* selects relevant attributes (including their possible combinations) automatically.

Unfortunately, space constraints do not permit a discussion in sufficient details of all types and principles of different networks – just a description of one of them would from the material for several books. If a reader needs a deeper understanding of these algorithms, he or she can learn the details using the literature in the bibliography. Here, only one of the options using probably the most famous, basic type of artificial neural networks – feed forward network – will be briefly demonstrated. A good introduction to various types and principles of artificial neural networks can be found, inter alia, in [113].

An artificial multi-layer neural network can also be viewed as a tool that can, in principle, learn hierarchically in the sense that it gradually creates attributes at higher and higher levels from attributes at lower levels, such as *character* → *syllable* → *word* → *phrase* → *sentence* → *paragraph* → *chapter* → ... to attributes at higher levels based on the number of levels (defined by the user), their composition (the number and type of neurons), and the type of connection from a lower to higher level (full or sparse to reduce attributes).

Many different applications, including text mining, have already proven the great benefits of deep learning. However, it is necessary to be aware that in general, artificial neural networks count among the algorithms of the group called black-box, so it is very difficult or (almost) impossible to prove excellent classification (or regression) results by sufficient explanation, not only that "it is so," but especially "why it is so."

During their training, neural networks form very abstract spaces with virtually any number of artificial dimensions that may substantially differ from the original input dimensions. If the proper classification is sufficient for the analysis and further investigation and synthesis can be done based just on the pure classification results, no problem may arise. But if there is a need for a different justification for the results 'because the neural network has said that', then even only classifying the correct classes may not fully satisfy.

For example, in medicine, it is not enough to just diagnose – it must be justified for further correct treatment. And the same can often apply to text mining as well: Which words and phrases are relevant to a particular category? Why are the same words or phrases in several different categories? Why is something irony somewhere and not elsewhere even if the irony is almost perfectly recognized? For black-box algorithms, a wide field is still open to investigate how to justify the knowledge found.

## 11.2 Artificial Neural Networks

An artificial neuron can be viewed as a transfer function $\phi(.)$ having $n+1$ inputs $x_i$ weighted by respective weights $w_i$, where $w_i \in \mathbb{R}, -\infty < w_i < +\infty$:

$$\hat{y} = \phi(\mathbf{w} \cdot \mathbf{x}) = \phi(\sum_{i=0}^{n} w_i x_i) = \phi(\theta + \sum_{i=1}^{n} w_i x_i), \qquad (11.1)$$

where for the dot product of the vectors $\mathbf{w} \cdot \mathbf{x}$ with the constant value of $x_0 = 1.0$ there is the bias (threshold) $\theta = w_0 x_0 = w_0 \cdot 1.0$ that determines the value of the transfer function $\phi(.)$ from which the neuron is activated and provides its output $\neq 0.0$. Hence, the neuron generally makes the transformation $\mathbb{R}^n \to \mathbb{R}$.

Determining the specific $\phi$ function depends on the type of application and known or anticipated properties of the approximated unknown function $\hat{y} = \hat{f}(\mathbf{x})$. A commonly used nonlinear transfer function $\phi$ is *sigmoid* (known also as *logistic*), $S(x)$, where

$$S(x) = (1 + e^{-1}), \lim_{x \to -\infty} S(x) = 0, \lim_{x \to +\infty} S(x) = 1,$$

and where, for classification, the usual threshold when the neuron returns 1 is for $S(x) \geq 0.5$, otherwise 0. However, there are also a number of alternatives required usually by specific applications for certain reasons; details can be found in literature.

Neurons can be combined in layers into a network to approximate unknown complex nonlinear (including discontinuous or non-differentiable) functions, which can also be used advantageously for classification or regression. The error function $E$ of the approximation (or classification in our case) is usually defined as follows:

$$E = \frac{1}{2}(y - \hat{y})^2, \qquad (11.2)$$

where $y$ is a known correct value and $\hat{y}$ is a value returned for the given training data sample by the trained network. The square of the difference $y - \hat{y}$ eliminates negative error values and highlights larger errors whose removal takes precedence over smaller errors (multiplication by $\frac{1}{2}$ is just for formal reasons, so that the number 2 in the exponent may not be considered in the derivative when looking for the steepest descent of the error).

To correct the output of $\phi(\mathbf{w} \cdot \mathbf{x})$ for the input sample $\mathbf{x}$, only the weight vector $\mathbf{w}$ can be modified because the values of $\mathbf{x}$ are given. This modification affects the significance of the $n$ weighted individual components (attributes $x_i$) of the input vector $\mathbf{x} = (x_1, x_2, \dots, x_n)$, the combination of which determines the required output value of the approximated function $\hat{f}(\mathbf{x})$.

Of course, no weights $w_i$ are known before starting the training process. These weights always determine the instantaneous position of a point representing a certain iterative solution in an $n$-dimensional space.

The goal is ideally the optimal solution – in our case, the maximal classification accuracy (alternatively the minimum classification error), which is the global maximum of the desired function $\hat{f}(\mathbf{x}) \approx f\mathbf{x}$).

The initial $w_i$'s values define the starting point and because it is typically not known where to start from, it is only possible to randomly generate those coordinate values. In a case when there would be some initial, if limited, knowledge available, it would be possible to determine the starting subspace (or point) more precisely.

Due to possible error on the trained network's output, it is necessary to appropriately adjust the weights of the attributes. In each iteration step, it is commonly gradually done for the individual layers in the direction from the output to the input layer – therefore the term 'backup error propagation' is often used. All the necessary detail can be found in, for example, [113], or other existing literature.

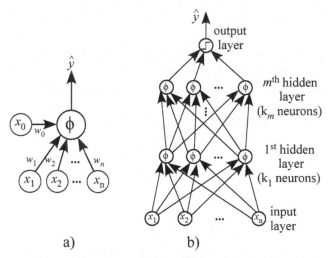

a)                              b)

**Figure 11.2:** An artificial neuron (a) and a multi-layer feed-forward classification neural network composed of neurons (thresholds are omitted) (b).

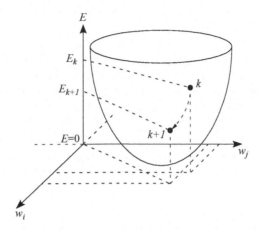

**Figure 11.3:** At an iteration step $k$, the steepest descent across the surface of the error function, $E$, is found and the coordinates (weights) of the point position are corrected so that it moves towards the lower error value at $k+1$.

In fact, it is about converting the original space described by the vectors **x** into a highly different abstract space described by the vectors **w**, which typically has many more artificially introduced dimensions (transformed attributes). In this new space, the neural network is looking for (and often successfully finds) the desired approximated solution.

Note that generally non-linear functions may suffer from having a lot of local extremes where it is easy to get stuck in finding the global extreme. If the idea of searching for the highest mountain in the misty, heavily mountainous landscape is used, then it is obvious that success can depend on the location from which the search begins – far from or near the foot of the highest mountain.

Therefore, it is advisable to search repeatedly from different starting points and compare the achieved results, which can be done by a newly generated set of the coordinates using different random-generator seeds.

If these results are the same or very similar, then it is very probable that the best achievable solution was found.

## 11.3   Deep Learning in R

Today, a lot of freely accessible software tools implementing various artificial neural networks are available. The tools commonly include many useful acceleration and optimization enhancements from both the mathematical and implementation point of view. In addition, there are still upcoming new enhancements or new types of neural networks, so the selection is fairly wide and the user can also choose from the task type being solved.

The R-language add-on system is no exception, and a variety of methods and algorithms can be found, sometimes more general, sometimes tailor-made for specific task groups. For ease of use and installation, this chapter confines itself to demonstrating the use of a single artificial neural network system – SNNS, which stands for Stuttgart Neural Network Simulator, one of the outstanding simulation systems [287].

The SNNS is called as a function `mlp()` with the main parameters `size=c(1000, 500, 200, 80, 20)` defining gradually the individual layer sizes from the network input to its output, and `maxit=50` defining the number of iterations.

```
library(Rcpp)
library(RSNNS)

# loading the data
data <- read.csv("BookReviews.csv", header=TRUE)

# converting the data to a matrix
dtm <- as.matrix(data)

# removing the object data (unnecessary from now)
rm(data)

# number of instances
n <- dim(dtm)[1]

# number of attributes
m <- dim(dtm)[2]

# indexes of 100 testing samples
number_of_test_samples <- 100
test_index <- 1:number_of_test_samples

# selecting training samples
train_data <- dtm[-test_index, 1:(m-1)]
train_labels <- dtm[-test_index, m]

# selecting testing samples
test_data <- dtm[test_index, 1:(m-1)]
test_labels <- dtm[test_index, m]
# training a neural network model
```

```
rsnns_model <- mlp(train_data, train_labels,
                   size=c(1000, 500, 200, 80, 20),
                   learnFuncParams=c(0.1),
                   maxit=50)

# plotting iterative errors of the rsnns_model object
plotIterativeError(rsnns_model, main="Iterative error")

# plotting a ROC curve for training data
plotROC(fitted.values(rsnns_model),
        train_labels,
        main="ROC curve - training data",
        xlab="1 - specificity",
        ylab="sensitivity")

# calculating class predictions for test data
predictions <- predict(rsnns_model, test_data)

# plotting a ROC curve for test data
plotROC(predictions,
        test_labels,
        main="ROC curve - test data",
        xlab="1 - specificity",
        ylab="sensitivity")

# creating a confusion matrix for training data (label 1
# is assigned when the output value > 0.6 and the remaining
# outputs are less than 0.4)
print(confusionMatrix(
        train_labels,
        encodeClassLabels(fitted.values(rsnns_model),
                          method="402040",
                          l=0.4, h=0.6)))

# transforming real numbers to 0 or 1
predictions <- round(predictions)

# creating a confusion matrix
cm <- table(test_labels, predictions)
print("Confusion matrix")
print(cm)
```

```
# number of instances
n <- sum(cm)

# number of correctly classified instances for each class
correct <- diag(cm)

# numbers of instances in each class
instances_in_classes <- apply(cm, 1, sum)

# numbers of each class predictions
class_predictions <- apply(cm, 2, sum)

# accuracy
accuracy <- sum(correct)/n

# precision per class
precision <- correct/class_predictions

# recall per class
recall <- correct/instances_in_classes

# F1-measure per class
f1 <- 2 * precision * recall / (precision + recall)

# printing summary information for all classes
df <- data.frame(precision, recall, f1)

print("Detailed classification metrics")
print(df)
print(paste("Accuracy:", accuracy))

# macroaveraging
print("Macro-averaged metrics")
print(colMeans(df))

# microaveraging
print("Micro-averaged metrics")
print(apply(df, 2, function (x)
                    weighted.mean(x, w=instances_in_classes)))
```

For the demonstrated classification example, the console gives the following textual output that contains the confusion matrix for training samples, a confusion matrix for testing samples, and the classification performance values for positive, negative, and all training samples:

```
         predictions
targets   0    1
      0  434    9
      1    2  455
```

```
[1] "Confusion matrix"
          predictions
test_labels  0   1
          0  40  17
          1   4  39
```

```
[1] "Detailed classification metrics"
   precision    recall          f1
0 0.9090909 0.7017544 0.7920792
1 0.6964286 0.9069767 0.7878788
[1] "Accuracy: 0.79"
```

```
[1] "Macro-averaged metrics"
precision    recall          f1
0.8027597 0.8043656 0.7899790
```

```
[1] "Micro-averaged metrics"
precision    recall          f1
0.8176461 0.7900000 0.7902730
```

In addition, three graphs are available. The first grapf, Figure 11.4, represents the decrease of the sum of squared errors $(y - \hat{y})^2$, SSE, during the iteration process. It clearly illustrates how the network learns.

The next graph, Figure 11.5, shows the dependence of the so called *sensitivity* value on the *specificity* value for the training data, while the third graph, Figure 11.6, shows the same for the test data. *Sensitivity* (or *recall*) is also known as *true positive rate*, $TP/(TP+FN)$, while specificity is known as *true negative rate*, $TN/(FP+TN)$.

Alternatively, this dependence can be also used to express the relative relationship between the positive examples mistaken for negative and negative ones mistaken for positive. Such a dependence function may be useful for visual comparison of classifier behaviour on test and training examples. The presented neural network learned the training samples almost perfectly for both classes while the testing samples gave more errors. However, it can be usually expected.

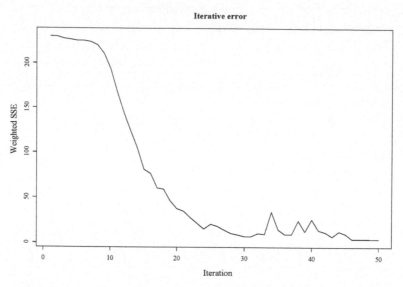

**Figure 11.4:** The squared sum of error decrease depending on an iteration step number during learning.

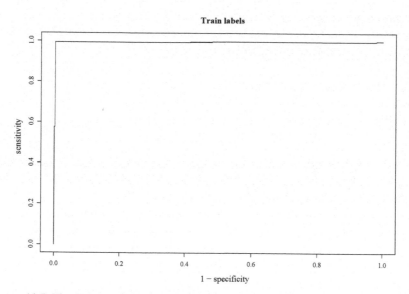

**Figure 11.5:** The dependence between the positive samples mistaken for negative and negative ones mistaken for positive (training samples).

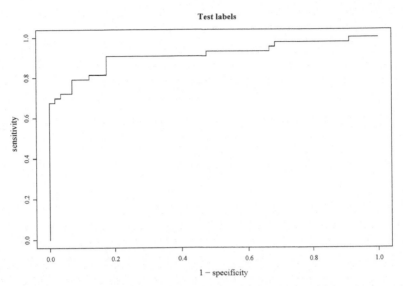

**Figure 11.6:** The dependence between the positive samples mistaken for negative and negative ones mistaken for positive (test samples).

The applied software package, *RSNNS*, is only one possibility for deep learning. There are also other popular packages that can be applied. One of the popular alternative packages which was originally developed for another programming language, *Python*, is known under the name *keras*. However, the machine learning community also enabled the use of *keras* for R; thus, it is freely available in R as well, including an R interface.

The referenced *keras* system is very rich in terms of usable features and parameters, so it is worth paying attention to the appropriate accessible documentation available, to a large extent (in various forms, with different content) via the Internet - unfortunately, space restrictions do not allow for more details here. An interested user may start from [50, 10, 83] and continue with other references offered in available resources.

# Chapter 12

# Clustering

## 12.1 Introduction to Clustering

Supervised learning which is the most common type of learning problem [71] is also very popular in text mining [245]. In order to successfully solve the problems belonging to the supervised learning category, the texts to be processed need to have assigned labels that describe the categories in the data. A supervised learning algorithm can, using the documents' properties and labels, generalize what documents in categories look like, what is typical for them, and how to correctly assign unlabeled instances to one or more existing categories.

Unfortunately, unavailability of the labelled data is often a major problem in text mining [228]. It is because the labeling process itself is very demanding. Reading tens or hundreds of documents and correctly assigning labels to them requires the effort of many people for many hours. Even when the human annotators are experts in the given field, the resulting quality of the labeling process is not have to always be high. It is not uncommon that a complete agreement of the annotators is not achieved [239]. Additionally, new data constantly occurs, so it is nearly impossible to have the labels assigned to the data in a reasonable time and in reasonable amounts.

*Clustering* is the most common form of *unsupervised learning* and enables automatic grouping of unlabeled documents into subsets called *clusters*. It is expected that the clusters are homogeneous internally and can be clearly distinguished from the others. This means that the documents contained in one cluster are more similar (e.g., share the same topic) to the remaining documents in the same cluster than to the documents in the other clusters. Here, we might anticipate that the document properties together with the concept of similarity play a crucial role in the clustering process.

Clustering has gained in importance with the increasing amounts of data available. It covers a wide variety of algorithms rooted in statistics, computer science, computational intelligence, and machine learning with applications in many disciplines [285].

Clustering has frequently been used successfully for organizing and searching large text collections, for example, in automatic creation of ontologies, summarizing, disambiguating, and navigating the results retrieved by a search engine, patent analysis, detecting crime patterns and many others [61, 70, 262, 108, 40]. If the clusters are good and reliable with respect to the given goal, they can be successfully used as classes or in order to improve classification [84, 158].

## 12.2  Difficulties of Clustering

In supervised learning tasks, the goal is usually well defined and described. For example, in a spam filtering problem, we want to separate spam e-mails from the relevant ones. What spam e-mails look like can be derived from a collection of good examples of spam. Then, a learning algorithm looks at the properties of spam e-mails and compares them with the properties of other e-mails. The relevant properties that can be used in distinguishing spam from the rest are somehow combined in the model (e.g., in the tests contained in the nodes of a decision tree) or in the measures quantifying their relationship to a category are determined (e.g., the probabilities in the naïve Bayes classifier). The success of the generalization can be measured by clearly defined measures like the correctness of classification. Unfortunately, one clear answer to a clustering problem and measures that indisputably evaluate the outcomes of unsupervised learning tasks do not exist.

Imagine an example from the clothing industry. We have information about the physical characteristics of our customers and we want to determine the suitable sizes of the clothes, take T-shirts, for instance. Since we have no information about the sizes to be made, we want to find groups of people with similar physical characteristics and to offer one size for each group. As the people in one group are expected to be similar, we hope the size will fit all of them. The question is, how many groups should we define? A larger number of available sizes will be appreciated by the customers but will put additional requirements on our business (developing more sizes, manufacturing more different sizes, keeping more sizes in stock, etc.). Having a lower number of models will make the business easier but can discourage the customers who do not find the perfect size. One extreme is to make only one size and provide it to all, the other possibility is to manufacture a unique size for each customer. All mentioned options are good from some perspective. At the same time, all of them are not good for some other reasons. It is therefore obvious that no single correct answer exists and that it depends on the problem that is to be solved and how the problem is evaluated. When emphasizing the convenience of the customers, more sizes should be provided. When the ease of the business has a higher importance, a lower number of T-shirt sizes would be preferred.

We might relate similar considerations to text documents; for example, newspaper articles. Clustering can be used in order to find groups of similar articles, sharing a common topic. This may be useful in an information retrieval task when relevant documents matching a user defined query should be returned. Clustering can be applied to the documents before retrieval, so groups of similar documents are found. These groups are represented by their centroids and every search request is first compared to the centroids. Subsequently, matching the query against documents is carried out only in the selected, highly relevant groups [234]. A question similar to the one on the T-shirts example arises here: How many groups of documents should be created?

Having more clusters means that they are more specific. For example, instead of one group containing documents related to sports, we may have more groups focused on football, ice hockey, athletics, the Winter Olympic Games, The National Hockey League, etc. More clusters will probably result in a more efficient way of retrieving the documents because a smaller number of documents needs to be compared to the query. The precision, i.e., the percentage of relevant documents, will most likely be higher because the clusters should contain documents that are somewhat similar and thus sharing a common topic. On the other hand, the recall, measuring the portion of relevant documents that have been retrieved, could be lower because some other clusters might also contain relevant documents, but other topics are prevailing there.

When the created clusters are to be used for subsequent classification we have to realize that it will be easier to distinguish documents about sports from other categories, the economy, politics, or nature, for instance. It will be more complicated to differentiate between documents about tennis, badminton, and table tennis, representing quite similar topics under the sports category.

Sometimes, information about the relationships between clusters, for example, knowing that groups of documents talking about athletics and football belong to another group talking about sports, might be interesting too. This means that a clustering technique that is able to produce a hierarchy of clusters is suitable in this situation.

It is obvious that there might be many goals and evaluation viewpoints. Thus, interesting structures in the document collection are expected to be found [247] and the evaluation must be often done by a human expert. There are also many measures which can be considered in the evaluation process that somehow quantify the properties of the created clusters (i.e., measure the quality of clustering).

## 12.3 Similarity Measures

Document similarity is the only endogenous information that is available in the clustering process. The similarity can be measured between individual documents, between groups of documents, or between a document and a group of documents. It is therefore necessary to define how to represent a group of documents. One of the possibilities is to represent a group of documents using its *centroid* (i.e., the average document):

$$Centroid_C = \frac{\sum\limits_{x \in C} x}{|C|}, \tag{12.1}$$

where $|C|$ is the number of documents in cluster $C$. A cluster can also be represented by one of the documents contained in it, for example, by its *medoid* (a document closest to the centroid). All this depends on the applied clustering algorithm. For instance, some algorithms measure the similarity between two clusters in terms of the similarity of the most similar objects (single linkage clustering method) or two most dissimilar objects (complete linkage clustering method). In the k-means algorithm, the objects are moved to the closest cluster which is determined according to the proximity to a cluster centroid. In agglomerative centroid methods, each cluster is represented by a centroid and the similarity is calculated between these centroids [230, 128].

Let a measure $d(x_1, x_2)$ represent the distance between objects $x_1$ and $x_2$. If it should be considered a correct metric, the following criteria must be fulfilled [242, 283, 250]:

■ $d(x_1, x_2) \geq 0$, i.e., the distance is always non-negative;

■ $d(x_1, x_2) = 0$ if $x_1$ and $x_2$ are identical which is known as self-proximity;

■ $d(x_1, x_2) = d(x_2, x_1)$ which means that the distance must be symmetrical;

■ $d(x_1, x_2) \leq d(x_1, x_3) + d(x_3, x_2)$ which is known as the triangle inequality.

If we look, for example, at the cosine similarity measure we can see that some of the conditions are violated. Thus, similarities are sometimes converted to distances using some simple monotonic function such as the linear transform or the inverse transform [130]:

$$Sim_{linear}(x_1, x_2) = C - Dist(x_1, x_2) \tag{12.2}$$

$$Sim_{inverse}(x_1, x_2) = \frac{1}{Dist(x_1, x_2) + C}, \tag{12.3}$$

where $C$ is a constant that enables the fitting of the similarity values to a desired interval.

Although the terms *similarity* and *distance* are sometimes used interchangeably, there is a difference between them. If, for example, two points in the space are close to each other, their distance is low, but their similarity is high. If the two points are identical, their distance is minimal (zero) and their similarity is maximal (typically 1). Thus saying, for example, that the Euclidean distance is used as a similarity measure or that the cosine similarity measures the distance between two vectors is not completely correct.

There are many possibilities of measuring the similarity between documents. They typically consider the values of the attributes of the compared documents and combine the outcome of the comparison into a single value. This value reflects the positions of the documents in the multidimensional space. The position can be perceived as single points or as a vector with the beginning in the origin of the space (the point $(0,0,0,...,0)$) and ending in that point. Generally, the closer the points or vectors appear, the more similar the documents are. Given examples of what is similar and what is not, it is also possible to learn a new distance metric that takes these requirements into consideration [283].

## 12.3.1 Cosine Similarity

In text mining tasks, *cosine similarity* based on the angle $\theta$ between vector pairs is very popular [138]. It is calculated as the cosine of the angle $\theta$ between two vectors $x_1$ and $x_2$:

$$d_{cosine}(x_1,x_2) = cos(\theta) = \frac{x_1 \cdot x_2}{||x_1||\,||x_2||} = \frac{\sum_{i=1}^{m} x_1^i \cdot x_2^i}{\sqrt{\sum_{i=1}^{m}(x_1^i)^2} \cdot \sqrt{\sum_{i=1}^{m}(x_2^i)^2}}, \quad (12.4)$$

where $x_1 \cdot x_2$ is the dot (linear, scalar) product of vectors $x_1$ and $x_2$ and $||x_i||$ is the length of vector $x_i$. The dot product has a favourable property which makes it popular for calculating similarities. If the values of $i^{th}$ attribute of $x_1$ and $x_2$ are not zeros (i.e., the documents share the same property, e.g., a word) the value of the dot product will be bigger. If one document has a certain property ($x_1^i > 0$) and the other not ($x_2^i = 0$) the value of the dot product will not increase because $x_1^i \cdot x_2^i = 0$.

The problem of the dot product is that longer vectors, i.e., vectors containing more values, cause the dot product to be higher. When the vectors are normalized to a unit vector (a vector with size 1) their dot product is then equal to the cosine of the angle between them. When the vectors are already normalized (we can achieve this by dividing the vectors by their lengths), the cosine similarity can be simply calculated as

$$d_{cosine}(x_1,x_2) = x_1 \cdot x_2 \quad (12.5)$$

The values of the cosine similarity range from 0 to 1 as the angle between the vectors can be between 0 and 90 degrees (we consider only that part of the vectors space where the attribute values are positive because the weights in document vectors are always positive – a document cannot contain a certain word a negative number of times). The more similar the documents are, the smaller the angle between their vectors is, so the value of this measure is higher.

The cosine similarity does not satisfy the conditions of being a metric. This can be easily solved by working with the angle between two vectors, not with its cosine [242]:

$$d_{cosine}(x_1, x_2) = \arccos \frac{\sum_{i=1}^{m} x_1^i \cdot x_2^i}{\sqrt{\sum_{i=1}^{m} (x_1^i)^2} \cdot \sqrt{\sum_{i=1}^{m} (x_2^i)^2}} \quad (12.6)$$

The *arccos* function (a function inverse to *cos*) gives us the angle from its cosine. The angle can be between 0 and $\pi/2$ so if we want to normalize the value of the distance measure (to keep the values between 0 and 1), we can simply divide it by $\pi/2$. To avoid complicated calculations of the *arccos* function, we can calculate the cosine distance as $1 - similarity$ [91].

## 12.3.2   Euclidean Distance

The *Euclidean distance* is a standard geometric distance measuring the distance of two points in an n-dimensional space (in a two or three dimensional space, it can easily be measured by a ruler). It is the implicit distance for the k-means algorithm [120].

$$d_{Euclidean}(x_1, x_2) = \sqrt{\sum_{i=1}^{m} (x_1^i - x_2^i)^2} \quad (12.7)$$

## 12.3.3   Manhattan Distance

*Manhattan distance* is also known as city block distance, according to the distance you need to walk between two places using only orthogonal streets (the name comes from the island of Manhattan where the streets are mostly organized into a rectangular grid). The distance is simply calculated as the sum of differences across all dimensions [97]:

$$d_{Manhattan}(x_1, x_2) = \sum_{i=1}^{m} |x_1^i - x_2^i| \quad (12.8)$$

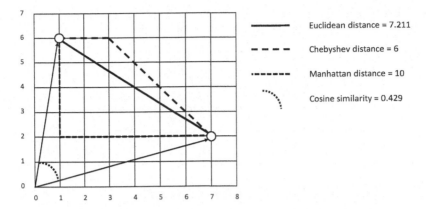

**Figure 12.1:** A graphical representation of selected similarity measures.

## 12.3.4 Chebyshev Distance

*Chebyshev distance* is also known as maximum distance or chessboard distance, according to the number of moves a king on a chessboard must make when moving from one place to another in the shortest way. It is calculated as the maximum from differences across all dimensions [284]:

$$d_{Chebyshev}(x_1,x_2) = \max_{i=1}^{m} |x_1^i - x_2^i| \tag{12.9}$$

## 12.3.5 Minkowski Distance

This distance is a generalization including the Euclidean, Manhattan, and Chebyshev distances. The specific form of the distance depends on the parameter $p$ known as the order:

$$d_{Minkowski}(x_1,x_2) = \sqrt[p]{\sum_{i=1}^{m}(x_1^i - x_2^i)^p}, \; p \geq 1 \tag{12.10}$$

If $p$ is 1, we have the Manhattan distance, $p = 2$ gives us the Euclidean distance, and $p = \infty$ represents the Chebyshev distance [16].

## 12.3.6 Jaccard Coefficient

The *Jaccard coefficient* measures the similarity in terms of intersection of two sets divided by their union. For text documents, the Jaccard coefficient compares the sum of weights of shared terms to the sum of weights of terms that are present in either of the two documents but are not the shared terms [120]:

$$
\begin{aligned}
d_{Jaccard}(d_1, d_2) &= \frac{d_1 \cdot d_2}{||d_1|| + ||d_2|| - d_1 \cdot d_2} \\
&= \frac{\sum_{i=1}^{m} d_1^i \cdot d_2^i}{\sqrt{\sum_{i=1}^{m} (d_1^i)^2} + \sqrt{\sum_{i=1}^{m} (d_2^i)^2} - \sum_{i=1}^{m} d_1^i \cdot d_2^i}
\end{aligned}
\tag{12.11}
$$

## 12.4 Types of Clustering Algorithms

We have a set, $X = \{x_1, x_2, \cdots, x_n\}$, of documents to be clustered, as the input. There are various algorithms which have a common goal – to assign the documents to appropriate clusters. The methods generally differ in the way of representing the similarities between documents and finding the clusters. The algorithms might be based on distances, probabilities, density, and other measures. The earliest methods of document clustering, like the k-means algorithm and agglomerative methods, were based on distances [5].

When an algorithm assigns each object to exactly one cluster, we talk about *hard clustering*. It is also possible to have each object associated with each cluster with a degree of membership in the interval $[0, 1]$. This approach is known as *soft* or *fuzzy clustering* [87]. Sometimes, a document does not have to be assigned to a cluster. This is called *non-exhaustive clustering* [277].

Clustering algorithms usually operate on two types of structures. The first type represents the objects to be clustered as sequences of objects' attributes (variables, features) in the form of vectors $x_i = (x_i^1, x_i^2, \cdots, x_i^m)$. All objects then form an object-by-variable matrix (e.g., a term-document matrix). The second structure is a collection of proximities that have to be available for all pairs of the objects. The proximities are then represented by a square object-to-object matrix containing similarities or dissimilarities of the objects [142].

### 12.4.1 Partitional (Flat) Clustering

The goal of *partitional (flat) clustering* is to find a partition $C = \{C_1, C_2, \cdots, C_k\}$, $k \le n$ of $X$ such that every document belongs to one and only one set of $C$ that are called *clusters*. This means, that the clusters are not overlapping and their union gives us the set $X$. Partitioning clustering can be formally described as follows [285]:

$$C_i \neq \emptyset, \quad i = 1, 2, \cdots, k$$

$$\bigcup_{i=1}^{k} C_i = X \tag{12.12}$$

$$C_i \cap C_j = \emptyset, \quad i, j = 1, 2, \cdots, k \text{ and } i \neq j$$

During flat clustering, the value of an *objective function* that quantifies the quality of a clustering solution is optimized. The objective function usually describes how similar the documents and the cluster centroids are, how dissimilar the centroids are, and so on; see below.

It is not computationally feasible to use brute force to examine all possible clustering solutions to find the best one. Thus, flat clustering usually starts with an initial solution based on some random element (the seed) and this solution is further improved in several iterations. When the initial solution is not well selected, the final partitioning does not have to be optimal (the search process does not reach the global optimum). Thus, some methods of pre-computing the seed to increase the chance of reaching the global optimum might be employed.

## 12.4.2 Hierarchical Clustering

*Hierarchical clustering* constructs a tree-like structure (hierarchy) $H = \{H_1, H_2, \cdots, H_h\}$ of subsets $H_i \subseteq X$ such that any two sets $H_i$ and $H_j$ are either disjoint or one contains the other. The clusters at the bottom of this hierarchy contain only a single object from $X$ (singleton clusters). One of the subsets, the root of the hierarchy, contains all clustered objects. Mathematically hierarchical clustering is described as [197]:

$$H_i \subset H_j, H_j \subset H_i \text{ or } H_i \cap H_j = \emptyset \text{ for any } i, j = 1, 2, \cdots, h, \text{ and } j \neq i$$

$$\{x\} \in H \text{ for each } x \in X \tag{12.13}$$

$$X \in H$$

Two approaches can be applied when building the hierarchy of clusters – top-down in case of *divisive* hierarchical clustering and bottom-up in *agglomerative* hierarchical clustering. While agglomerative algorithms are strictly local, meaning that they consider only the closest neighbourhood of clusters, divisive algorithms are more global because the context of all data has been, in some way, used so far when partitioning a cluster into subsets [3].

The result of hierarchical clustering can be visualized using a *dendrogram* where the proximity of individual objects as well as the clusters is well visible. A dendrogram over a set $X$ is a binary rooted tree where the leaves correspond to

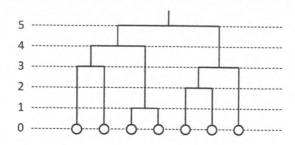

**Figure 12.2:** A dendrogram with levels [3].

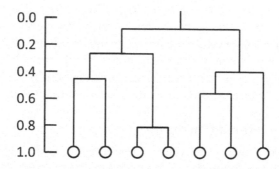

**Figure 12.3:** A dendrogram with combination similarities [185].

the elements of X. Every node is assigned a level using a level function. Leaves are placed at level 0, parents have higher levels than their children, and no level is empty. The objects contained in any cluster from the tree are the leaves of a subtree, starting in a node representing the cluster. The level function assigns zero to leaves and higher values to parent nodes; see Figure 12.2 [3].

Alternatively, the level of a cluster can be expressed by some similarity measure between the two clusters that are included in their union. This similarity is known as *combination similarity*. In case of using the cosine similarity, the leaves in a dendrogram have similarity 1 (a document is identical, i.e., the most similar with itself) and decreases towards the root; see Figure 12.3. In an agglomerative approach, the combination similarities of the successive merges are usually monotonic, i.e., $s_1 \leq s_2 \leq \cdots \leq s_N$. A situation where this condition is violated is called inversion, i.e., there exists at least one $s_i \geq s_{i+1}$. This may happen in the case of using centroid clustering [185].

Hierarchical clustering methods produce a complete arrangement of documents in clusters, including hierarchical relationships between clusters. Sometimes, a specific number of clusters might be needed as in case of flat clustering. It is therefore necessary to cut the dendrogram to define the clusters. There are several criteria that might be used [185, 292]:

■ The dendrogram is cut so the result is a desired number of clusters, $k$, that is given in advance. This is quite similar to flat clustering where $k$ needs to be known too.

■ The similarity of clusters is taken into consideration and the dendrogram is cut at a desired level. The higher the similarity needed inside of the clusters, the more the clusters created.

■ The dendrogram is cut where the difference between two successive levels of combination similarity or a clustering criterion function is the largest. Adding an additional cluster would significantly decrease the quality of clustering significantly so it makes sense to stop clustering here. It is a similar approach to the elbow method mentioned below.

Hierarchical clustering is more descriptive than predictive because it uses document features only indirectly when constructing a dendrogram. The dendrogram describes how the existing documents are assigned to clusters but does not specify any borders for the clusters. In case of the k-means algorithm, the Voronoi boundaries separating the feature space can be easily found, see Figure 12.4, and used, for example, to predict where an unknown abject will belong [89].

Hierarchical clustering will also always find a clustering solution even when a meaningful solution does not exist. For an excellent example, see [89, p. 256].

Agglomerative clustering methods can often lead to clustering of a low quality. The problem is that documents usually share a significant portion of words. This can be true even for documents from different classes representing, for example, different topics. Thus, two documents that are nearest neighbours (are most similar) are placed to the same cluster although they should not. Percentages of documents whose neighbours are from a different class for different data sets containing textual documents is reported by [256] (the percentage is typically about 20%). Such wrong cluster assignments happen in the early phases of clustering and cannot be fixed later.

## 12.4.3 Graph Based Clustering

The previously mentioned algorithms supposed the documents to be clustered were represented by vectors. Their similarity was the basis for clustering. There is, however, an alternative view about the relationship between documents. We can compute similarity values between all pairs of documents and represent them in a graph. Vertices represent the documents and the edges that exist between

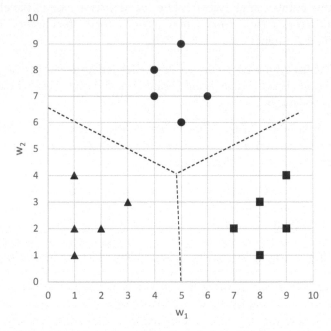

**Figure 12.4:** Voronoi boundaries for a clustering solution of the k-means algorithm partitioning objects into three clusters.

each pair of the vertices, have their weight equal to the similarity of the corresponding documents [46]. Another possibility is to represent the document set by a bipartite graph where the set of vertices consists of two sets, a set of documents and a set of terms. When a document $d_i$ contains a term $t_j$, an edge exists between these two vertices. Its weight is given by the tf-idf value of term $t_j$ in document $d_i$ [291].

The task of clustering documents in this representation is based on the same idea as in the case of other methods. The graph is partitioned into subgraphs so the weights of the links within clusters are high and the weights of the links between clusters are low.

## 12.5 Clustering Criterion Functions

During clustering, the best division of documents to clusters should be found. The quality of clustering can be measured by the function known as *criterion function* [292, 9] or clustering quality function [87]. Clustering can be thought of as an optimization problem because it tries to find a solution that maximizes or minimizes the value of the criterion function. For example, the function to be optimized (minimized) by the k-means algorithm is

$$CF_{k-means} = \sum_{i=1}^{k} \sum_{d \in C_i} similarity(d, c_i), \qquad (12.14)$$

where $k$ is the number of clusters and $c_i$ is the centroid of cluster $C_i$.

In some situations, the clustering criterion function is not needed because it is not feasible to evaluate all possible clustering solutions (for partitioning $n$ objects into $k$ clusters, $k^n/n!$ solutions would need to be examined). The algorithms are often greedy in nature and believe that using only a similarity measure to determine how much a document is similar to the objects in a cluster and dissimilar to documents in other clusters is enough [87].

Various clustering criterion functions were proposed, for example, by Zhao and Karypis [291]. Some of their functions are mentioned below (another set of criterion functions is presented by [9]). The functions were initially intended for partitioning algorithms but they can be used for hierarchical clustering as well. For example, instead of considering some typical merging criteria used during agglomerative clustering (see below), two clusters are merged so that the value of the criterion function improves the most. The criterion function is thus optimized only locally in a given stage of the algorithm [292].

## 12.5.1 Internal Criterion Functions

*Internal criterion functions* try to maximize the similarity of documents in individual clusters while documents in different clusters are not considered. The internal criterion function $I_1$ maximizes the sum of average pairwise similarities between the documents assigned to each cluster, weighted according to the size of each cluster. When the cosine similarity measure is used, the $I_1$ function is defined as follows:

$$I_1 = \sum_{c=1}^{k} n_c \left( \frac{1}{n_c^2} \sum_{x_i, x_j \in C_c} d_{cosine}(x_i, x_j) \right), \qquad (12.15)$$

where $k$ is the number of clusters and $n_c$ the number of documents in cluster $C_c$.

The internal criterion function $I_2$ maximizes the similarity between individual documents and the centroids $C_c$ of the clusters where the documents are assigned. The value of $I_2$ for the cosine similarity is calculated as

$$I_2 = \sum_{c=1}^{k} \sum_{x_i \in C_c} d_{cosine}(x_i, c_c), \qquad (12.16)$$

where $k$ is the number of clusters and $c_c$ is the centroid of cluster $C_c$.

## 12.5.2 External Criterion Function

*External criterion function* $E_1$ focuses on optimization of dissimilarity of individual clusters. Thus, using this function, the documents of each cluster are separated from the entire collection:

$$E_1 = \sum_{c=1}^{k} n_c \cos(c_c, c_{all}), \qquad (12.17)$$

where $k$ is the number of clusters, and $c_c$ is the centroid of cluster $C_c$, and $c_{all}$ is the centroid of the entire document collection. The value of $E_1$ is being minimized, which means that the angle of vectors $c_c$ and $c_{all}$ is the biggest possible.

## 12.5.3 Hybrid Criterion Functions

*Hybrid criterion functions* do not optimize only one criterion but a number of them. Instead of focusing only on intra-cluster similarity of the clustering solution while not taking into account documents in different clusters or focusing only on inter-cluster similarity and not considering similarity inside the clusters, hybrid functions combine both internal and external criteria. The two hybrid criterion functions combine the external function $E_1$ with $I_1$ or $I_2$ respectively:

$$H_1 = \frac{I_1}{E_1} \qquad (12.18)$$

$$H_2 = \frac{I_2}{E_1} \qquad (12.19)$$

Both internal functions $I_1$ and $I_2$ are normally maximized when the cosine similarity is used as the similarity measure. Although $E_1$ is normally minimized, hybrid functions $H_1$ and $H_2$ are maximized because their values are inversely related to $E_1$.

## 12.5.4 Graph Based Criterion Functions

A natural approach to graph-based clustering would be to find the minimal cut on the edges of the similarity graph $S$ (a graph where the vertices are the documents, there are edges between each pair of vertices, and the weights of the edges represent the similarity of the documents).

The *edge-cut* is defined as the sum of the edges connecting vertices representing documents from one cluster to vertices representing documents in other clusters:

$$\sum_{c=1}^{k} cut(S_c, S - S_c), \qquad (12.20)$$

where $S$ is a similarity graph representing a document collection, $S_c$ is a sub-graph representing cluster $c$, and $k$ is the number of clusters. Criterion function $G_1$, however, also considers the sum of the internal edges so it is not only the external perspective which is employed on clusters [73]:

$$\sum_{c=1}^{k} \frac{cut(S_c, S - S_c)}{\sum\limits_{x_i, x_j \in S_c} similarity(x_i, x_j)}, \tag{12.21}$$

When the cosine similarity measure is used, the $G_1$ criterion function is defined as

$$G_1 = \sum_{c=1}^{k} \frac{\sum\limits_{x_i \in S_c, x_j \in S - S_c} \cos(x_i, x_j)}{\sum\limits_{x_i, x_j \in S_c} \cos(x_i, x_j)} \tag{12.22}$$

Since both internal and external similarities are considered in this process, this criterion function $G_1$ might also be perceived as a hybrid one.

The normalized cut criterion function working on a bipartite graph (vertices represent documents and terms, edges correspond to the presence of terms in the documents) looks for an optimal edge-cut. The normalized cut [289] criterion function is defined as

$$G_2 = \sum_{c=1}^{k} \frac{cut(V_c, V - V_c)}{W(V_c)}, \tag{12.23}$$

where $V_c$ is the set of vertices assigned to the $c_{th}$ cluster, and $W(V_c)$ is the sum of the weights of the adjacency lists of the vertices assigned to the $c^{th}$ cluster ($c^{th}$ cluster contains both documents as well as terms).

## 12.6 Deciding on the Number of Clusters

The desired number of clusters is an important parameter, especially of flat clustering algorithms. The problem is that this number is not often known in advance because the true cluster structure is unclear. In the ideal case, only a human expert can say whether a clustering solution is suitable for a specific task. Sometimes, the number of clusters might be given by technological limitations, as in the case of the Scatter-Gather algorithm, where no more than 10 clusters could originally fit to a PC screen [185].

There are five clearly visible clusters on Figure 12.6. The objects were clustered to 1 to 10 clusters using the k means method using the Euclidean distance as the similarity measure. The within cluster sum of squared errors which is minimized during clustering was calculated for each of the solutions. How the value of this measure changes depending on the number of clusters can be seen in the

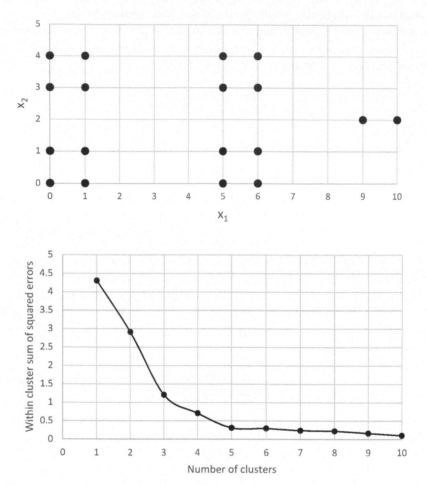

**Figure 12.5:** A demonstration of the elbow method to determine the number of clusters. Top: Objects to be clustered. Bottom: A dependency of clustering quality on the number of clusters. A clear 'elbow' is visible at 5 clusters; from this point, no significant improvement of the clustering quality occurs.

lower part of Figure 12.6. When each object is in a separate cluster, the sum of errors is zero, which is the global minimum. The clustering, however, is not useful.

We can see that, while the value of the error function decreases quite significantly, this sharp decrease stops when the number of clusters is five. From this point, the error decreases only slightly. We can see an *'elbow'* or a 'knee' on the curve approximating these values of the error. We can see that five clusters is a good compromise between the quality of clustering and the generality. Adding more clusters will not improve the solution considerably; having a smaller number of clusters, on the other hand, will increase the error significantly [229]. Similarly to the within cluster sum of squared errors, any other function measuring the quality clustering can be used.

Because the goal of clustering is to discover some natural division of data, it is expected that the objects are well located within clusters and separated from the other clusters. Different clustering solutions contain groups of objects with different levels of homogeneity and separation. A measure that is able to quantify both aspects can be used as the basis for selecting the right number of clusters too. For example, the *Silhouette method* (see below) is one of them. The average silhouette is calculated for different clustering solutions and a solution with the best value is chosen.

There are also other recommendations depending on a specific task. For example, in information retrieval, a formula to determine the right number of clusters was developed by Can and Ozkarahan [42]:

$$k = \frac{m \cdot n}{t}, \tag{12.24}$$

where $n$ is the number of terms, $m$ the number of documents, $t$ is the total number of term assignments.

## 12.7 K-Means

*K-means* is the most famous algorithm for flat clustering. It iteratively improves an initial clustering solution to minimize the distance between documents and centers (means) of the clusters where the documents belong by changing cluster assignments. The procedure can be described by the following steps:

1. $K$ documents (a seed) are initially selected as cluster centers $c_1, c_2, \cdots, c_k$.

2. For each document $d_i \in D$, its nearest center $c_j$ is found and the document is assigned to the corresponding cluster $C_j$ ($C_j \leftarrow C_j + \{d_i\}$).

3. The cluster centers are recomputed according to the current document assignment in clusters.

4. If a stopping criterion is not met, step 2 is revisited, otherwise a clustering solution is found.

There are a few commonly employed stopping criteria [185]:

■ The assignment of documents to clusters (i.e., the positions of centroids) does not change between iterations. This means that a global or local optimum has been found.

- The value of the criterion function reaches a certain threshold. This means that the clustering solution has a sufficient quality.

- The change (improvement) of the the criterion function value is below a certain threshold. This means that the clustering solution almost converged to an optimal solution (maybe to a local optimum).

- A certain number of iterations have been carried out. This approach does not ensure that the number of iterations is sufficient to have a clustering solution of a desired quality.

If an outlier is selected as a seed it is very likely that this document will remain alone in a cluster, so it is quite useless. Removing outliers might be a good method to prevent this problem.

Selection of a suitable seed is crucial for the success of flat clustering methods because different seeds might lead to different partitionings. Thus, the algorithm can be carried out several times with a different seed and the best solution can be chosen. Alternatively, the seed can be selected based on some external knowledge as a result of hierarchical clustering method, for example [257].

K-means produces convex (spherical) clusters and is not able to find clusters of other shapes. This might be a problem, especially when the clusters lie close to each other. This relates to another potential problem. The spherical shape of the produced clusters causes the cluster center to be shifted towards outliers, especially the ones placed very far [8]. To overcome the sensitivity to outliers, the k-medians [5] method, that uses medians instead of means for cluster representatives, can be applied.

## 12.8  K-Medoids

The *k-medoids* algorithm [7] works in a similar fashion as the k-means algorithm. A medoid is a document that is closest to the centroid. The benefit of using the medoids instead of the means is in the decrease of computational complexity. In order to compute a distance between two documents, all non zero elements of their vectors are used in the calculation. Compared to single documents that are usually represented by very sparse vectors, centroids in the form of the cluster means contain fewer non zero elements because they average the weights of all elements of the documents in the cluster they represent. Using the medoids instead of centroids thus decreases the number of calculations when calculating the distances.

## 12.9 Criterion Function Optimization

Zhao and Karypis [294] proposed a procedure that can be successfully used when clustering large data sets of documents. It contains an initial clustering and several iterations of cluster refinement. The goal of the refinement phase is to improve the value of a given *criterion function* (this algorithm may be used directly for functions $I1$, $I2$, $E1$, $H1$, $H2$, and $G1$; in the case of function $G2$, the algorithm is modified so that not only vertices representing documents, but also vertices representing terms are processed). The procedure can be expressed by the following steps:

1. Initially, $k$ documents are selected as the seeds.

2. Each document is assigned to a cluster according to the most similar document from the seed.

3. For each document (documents are visited in a random order), the following steps are carried out:

   ■ The value of the criterion function to be optimized is calculated.

   ■ The document is assigned to all other clusters and the value of the criterion function is calculated for each such assignment.

   ■ The document is moved to the cluster that improves the value of the criterion function the most. It is possible that the document remains in the same cluster when no improvement can be achieved.

4. If some documents have been moved, step 3 is repeated, otherwise end.

Compared to the k-means method, this method moves an object immediately after is it clear that it will improve the value of the criterion function. This is known as an incremental algorithm [294].

As in the case of the k-means algorithm, this greedy approach does not have to lead to an optimal solution and a local optimum of the criterion function can be found. To prevent this, the procedure can be repeated several times and the best solution is selected.

## 12.10 Agglomerative Hierarchical Clustering

In agglomerative hierarchical clustering, individual documents are repeatedly merged to form larger and larger clusters. This computationally demanding procedure has the following general steps:

1. All $n$ documents to be clustered are put to separate clusters (we therefore have $n$ clusters $C_1, C_2, \cdots, C_n$).

2. The similarity of each cluster to the remaining clusters is calculated. The similarities may be contained in a similarity matrix.

3. The maximal similarity is found so the two most similar clusters $C_i$ and $C_j$ can be identified.

4. Clusters $C_i$ and $C_j$ are merged, so they form a new cluster.

5. Steps 2, 3, and 4 are repeated until only one cluster remains.

A crucial aspect of the procedure which influences the nature of created clusters, is the measure of calculating the similarity between clusters. The similarity might be based on comparing one document from each cluster, cluster centers, or all possible pairs of documents from each cluster. All of the measures use some function $d$ (e.g., the cosine similarity) to calculate a specific value of similarity of two clusters to be subsequently merged. Commonly used methods for merging clusters follow [5, 185, 230, 128, 45, 81]:

■ *Single Linkage Clustering (SLINK)*: the clusters are merged according to the similarity of the most similar objects from the clusters. It emphasizes the regions where clusters are close, ignoring the overall shape of the original and newly created clusters. Another disadvantage is the sensitivity to outliers.

$$d_{SLINK}(H_i, H_j) = \min_{x_i \in H_i, x_j \in H_j} d(x_i, x_j) \tag{12.25}$$

When the newly created cluster is to be merged with a third cluster, the similarity can be calculated as follows:

$$d_{SLINK}(H_k, (H_i, H_j)) = \min(d(H_k, H_i), d(H_k, H_j)) \tag{12.26}$$

■ *Complete Linkage Clustering (CLINK)*: determines the similarity according to the similarity of two most dissimilar objects. In other words, the union of two clusters has the smallest diameter and the clusters tend to be more compact. Outliers again cause the problem – here the resulting cluster will be very big.

$$d_{CLINK}(H_i, H_j) = \max_{x_i \in H_i, x_j \in H_j} d(x_i, x_j) \tag{12.27}$$

When merging the new cluster with another one, the distance is calculated as:

$$d_{CLINK}(H_k, (H_i, H_j)) = \max(d(H_k, H_i), d(H_k, H_j)) \tag{12.28}$$

■ *Arithmetic average methods*: These methods calculate the distance between two clusters to be merged as an average distance between all cluster members. The *Unweighted Arithmetic Average Clustering* or *Unweighted Pair-Group Method using Arithmetic averages* (UPGMA) gives the same weight to all objects in the computation.

$$d_{UPGMA}(H_i, H_j) = \frac{1}{N_i \cdot N_j} \sum_{x_i \in H_i} \sum_{x_j \in H_j} d(x_i, x_j) \qquad (12.29)$$

When $H_i$ and $H_j$ are merged, the distance to a third cluster, $H_l$ can be calculated as follows:

$$d_{UPGMA}(H_k, (H_i, H_j)) = \frac{1}{2}(d(H_k, H_i) + d(H_k, H_j)) \qquad (12.30)$$

Information about the sizes of merged clusters is thus lost. In the *Weighted Pair-Group Method using Arithmetic averages (WPGMA)*, the distances are weighted according to the number of objects in each cluster (bigger clusters are down-weighted):

$$d_{WPGMA}(H_k, (H_i, H_j)) = \frac{N_i}{N_i + N_k} d(H_k, H_i) + \frac{N_j}{N_j + N_k} d(H_k, H_j) \quad (12.31)$$

■ *Centroid Agglomerative Clustering*: This method eliminates the computational complexity of the Group Average Agglomerative Clustering method because every cluster is represented by its mean and not by all objects. When replacing all objects by their mean (centroid), the monotonicity of the combination function can be violated leading to non-intuitive dendrograms (similarity of merged clusters is higher than similarity of some previously merged clusters). It is not necessary that this approach will work well for all data sets, but is is quite successful in document clustering.

The first method, known as *Unweighted Pair Group Method using Centroids (UPGMC)* treats all clusters as equally important, without considering their size:

$$d_{UPGMC}(H_i, H_j) = d(c_i, c_j), \qquad (12.32)$$

where $c_i$ and $c_j$ are the centroids of clusters $H_i$ and $H_j$.

When measuring the similarity to a third cluster, the following calculation is performed:

$$d_{UPGMC}(H_k, (H_i, H_j)) = \frac{N_i}{N_i + N_j} d(H_k, H_i)$$
$$+ \frac{N_j}{N_i + N_j} d(H_k, H_j) \qquad (12.33)$$
$$- \frac{N_i N_j}{(N_i + N_j)^2} d(H_i, H_j)$$

In contrast, the *Weighted Pair Group Method with Centroid (WPGMC)*, also known as median clustering, down-weights bigger clusters so the similarity is calculated independently of the number of objects in the clusters:

$$d_{WPGMC}(H_k, (H_i, H_j)) = \frac{d(H_k, H_i) + d(H_k, H_j)}{2} - \frac{d(H_i, H_j)}{4} \qquad (12.34)$$

■ *Ward Distance*: Ward distance is defined as the change in the sum of squared errors of a clustering solution when two clusters are merged. The errors are calculated as a squared distance of an element from its corresponding centroid. This measure is, in fact, the value that is minimized by the k-means algorithm.

$$d_{Ward}(H_1, H_2) = \sum_{x \in H_1 \cup H_2} \|x - c\|^2 - \sum_{x \in H_1} \|x - c_1\|^2 - \sum_{x \in H_2} \|x - c_2\|^2$$
$$= \frac{N_1 \cdot N_2}{N_1 + N_2} \cdot \|c_1 - c_2\|^2$$

$$(12.35)$$

where $N_i$ is the number of documents in cluster $i$, $c_i$ is the centroid of cluster $i$, and $c$ is the centroid of a cluster that is the union of the merged clusters [198].

To generalize the calculation of the distance calculation between cluster $k$ and a cluster that was created by merging clusters $i$ and $j$, the *Lance and Williams recurrence formula* [81] can be used:

$$d_{k(ij)} = \alpha_i d_{ki} + \alpha_j d_{kj} + \beta d_{ij} + \gamma |d_{ki} - d_{kj}| \qquad (12.36)$$

The values of parameters for the above mentioned methods can be found in Table 12.1.

**Table 12.1:** Parameters for the Lance and Williams recurrence formula [81, 246].

| Method | $\alpha_i$ | $\beta$ | $\gamma$ |
|---|---|---|---|
| SLINK | $\frac{1}{2}$ | 0 | $-\frac{1}{2}$ |
| CLINK | $\frac{1}{2}$ | 0 | $\frac{1}{2}$ |
| UPGMA | $\frac{N_i}{N_i+N_j}$ | 0 | 0 |
| WPGMA | $\frac{1}{2}$ | 0 | 0 |
| UPGMC | $\frac{N_i}{N_i+N_j}$ | $-\frac{N_iN_j}{(N_i+N_j)^2}$ | 0 |
| WPGMC | $\frac{1}{2}$ | $-\frac{1}{4}$ | 0 |
| Ward's | $\frac{N_k+N_i}{N_k+N_i+N_j}$ | $-\frac{N_k}{N_k+N_i+N_j}$ | 0 |
| Flexible[1] | $\frac{1}{2}(1-\beta)$ | $\beta < 1$ | 0 |

[1]$\alpha_1 + \alpha_2 + \beta = 1$

## 12.11  Scatter-Gather Algorithm

Flat methods are more efficient compared to the hierarchical methods. They are, however, dependent on the initial seed (e.g., initial cluster centers in the k-means method) so they are not as robust as the hierarchical algorithms that compare all pairs of documents. A solution is to combine both approaches to a hybrid method used in the *Scatter-Gather algorithm*. The Scater-Gather method is originally a document browsing method that partitions (scatters) a document collection to a small number of groups and presents their short summaries to a user. One or more groups are selected (gathered) by the user and these groups are further analyzed using the same approach (i.e., clustering, summary, and groups selection). By a repeated application of these scatter and gather steps, the gathered clusters are more and more detailed and closer to the user's information need [61].

Such an approach requires a fast clustering phase so a hybrid approach using a hierarchical method to select $k$ seeds to be used in a flat method was proposed in [61]. To select the initial seed, two approaches might be used:

■ *Buckshot*: Having $n$ documents to be clustered into $k$ clusters, $\sqrt{k \cdot n}$ documents are randomly selected and clustered using an agglomerative method to form $k$ clusters. These $k$ clusters then become the seeds for a flat method applied to the entire data set. Because the complexity of a hierarchical method is $O(n^2)$, using only about the square root of the documents, the complexity becomes $O(k \cdot n)$.

■ *Fractionation*: The documents are divided into $n/m$ buckets where $m > k$ is the size of each bucket. In each bucket, the documents are clustered using an agglomerative hierarchical method until the number of clusters in a bucket is reduced by a certain factor (initially, the number of clusters

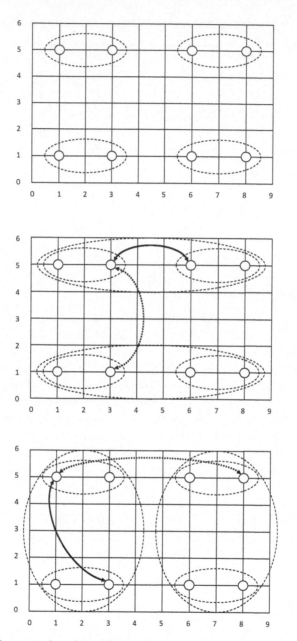

**Figure 12.6:** A demonstration of the difference between SLINK and CLINK methods. Top: An initial clustering solution. Middle: Newly created clusters using SLINK. The distance between the most similar objects represented by the solid line is smaller than the one represented by the dotted line. Bottom: Newly created clusters using CLINK. The distance between the most dissimilar objects represented by the solid line is smaller than the one represented by the dotted line.

is equal to the number of documents in a bucket). Subsequently, the process of rearranging the objects (documents or clusters of documents) into buckets and agglomerating them is repeated until there are $k$ seeds (roots of dendrograms obtained by the agglomeration procedure) found.

To further improve the seed selection results, the documents are not assigned to the buckets completely randomly, but they are first ordered according to the value of the index of the $i^{th}$ most frequent word in each document ($i$ is typically a smaller number, such as three, so medium frequency words are favoured). This ensures that documents that are close in this ordering share at least one word.

To increase the speed of computation of the buckshot phase, Liu et al. [180] use the initial $k$ centroids that are found in a pre-computed hierarchical agglomeration structure constructed in the first phase of their algorithm.

## 12.12 Divisive Hierarchical Clustering

Divisive clustering constructs the hierarchy in a top-down manner. All documents are initially in one cluster which is recursively divided until we have only singleton clusters. To obtain the same type of hierarchy as in case of agglomerative clustering, the cluster is usually divided into two sets.

The procedure of divisive clustering has the following general steps:

1. All $n$ documents to be clustered are put to one cluster.

2. The cluster is partitioned to $m$ clusters, usually using a flat method.

3. Step 2 is recursively applied to newly created clusters until every document is in a separate cluster.

The problem of divisive clustering is that having $n$ documents, they can be split into $2^{n-1} - 1$ possible non-empty groups. Even for a low $n$ there are so many possibilities that it is not computationally feasible to choose the best split [285].

In divisive clustering, is is often not necessary to build a complete dendrogram. At times, it may be useful to set some termination criteria for splitting the clusters to get a meaningful result, also because the clustered data can contain noise. Thus, particularly when no complete hierarchy needs to be created, the divisive clustering procedure is more efficient than the agglomerative [222]. The stopping criteria may include [53]:

■ $k$ clusters are found,

■ maximal hierarchy depth is reached,

■ minimum number of documents in a cluster remain,

■ minimal desired homogeneity (dissimilarity) in a cluster is achieved.

To get a hierarchy of clusters, a flat method is often recursively applied to the clusters being created. The flat methods that are usually used to split a cluster into more clusters require a number of clusters as an input. Thus, splitting a cluster into two (bi-partitioning) is a common approach. In this way, the created hierarchical structure has the same form as the one created in an agglomerative procedure when two clusters are merged [53]. If the same type of dendrogram that is a result of an agglomerative method is to be obtained, a sequence of $n-1$ ($n$ is the number of documents) bisections should be applied to the data (bisection means splitting a set into two subsets) [293].

Other possibilities of creating a cluster hierarchy include trying all possibilities of splits and evaluating them using the same metrics as in the case of agglomerative methods [232, 55]. However, because complete information about all objects is initially available, such divisive clustering can be considered to be a global approach [222] unlike agglomerative clustering.

In the divisive approach, the question is, which of the clusters to split. There are a few possible approaches [294, 240]:

■ In the simplest approach, the larger cluster is divided. In many cases, this leads to reasonably good and balanced clustering solutions. However, this approach is not suitable when naturally large clusters exist, because these large clusters will be divided first.

■ One of the current clusters is split so the value of a selected criterion function improves the most. Often, the cluster with the largest variance (i.e., the average squared distance to the cluster centroid) is selected, which is in fact optimization of the internal criterion function.

An example of the divisive clustering method is the bisecting k-means algorithm. The documents forming one cluster are split into two clusters using the basic k-means algorithm. Then, the cluster with a higher number of objects or with the largest variance is selected and split again. The selection and splitting steps are repeated until a desired number of clusters is reached or when a complete hierarchy is created [256]. It has been found in the text mining domain, that the bisecting k-means is generally better for higher numbers of clusters than the direct k-means method [141, 294].

## 12.13 Constrained Clustering

Agglomerative hierarchical clustering is strictly local and does not take the properties of the entire data set into consideration. This might be a disadvantage in some situations, especially when clustering documents share a large portion of their vocabulary. To prevent such a problem, an agglomerative clustering procedure using constraints imposed by a flat clustering is an option. Initially, the entire collection is partitioned to $k$ constraint clusters and clustering then takes place only within these clusters. It is, therefore, not possible to merge documents or clusters from different constraint clusters. This approach combines the global view of flat methods and local view of agglomerative methods. The time complexity of the agglomerative algorithm is also decreased [294].

Another possibility of change in the unsupervised process of clustering is to bring some external knowledge in the form of *must-links* (information about objects that must be in the same cluster) and *cannot-links* (information about objects that cannot be in the same cluster) [18]. The COP-K-means algorithm proposed by [272] then simply modifies the steps of the traditional k-means algorithm:

1. $K$ documents (a seed) are initially selected as cluster centers $c_1, c_2, \cdots, c_k$.

2. For each document $d_i \in D$, its nearest center $c_j$ *that does not violate any constraint* is found. Subsequently, the document is assigned to the corresponding cluster $C_j$ ($C_j \leftarrow C_j + \{d_i\}$). *If no cluster satisfying the constraints is found, the procedure fails.*

3. According to the current document assignment in clusters, the cluster centers are recomputed.

4. If a stopping criterion is not met, go to step 2, otherwise a clustering solution is found.

## 12.14    Evaluating Clustering Results

The general goal of clustering is to find some interesting structure in data. As there is usually no prior knowledge about the desired outcome and several alternative results are possible, evaluating clustering is not an easy task. Clustering also should not be considered an isolated mathematical problem but should always be studied in the context of a given task where it is employed [268, 115]. Besides using some mathematical computations described in this section, the suitability or usefulness of clustering for a specific application can be measured very naturally. For example, [185] use the time needed to find a desired answer in an information retrieval task that used clustering to navigate search results.

The goal of clustering evaluation is often to compare different clustering experiments and results. It is therefore desirable to use a simple numeric measure that is not biased toward a specific number and sizes of clusters and similarity measures, one that can easily be interpreted and can be used for a single cluster as well as for the entire clustering solution [244].

*Internal* (or intrinsic) measures evaluate clustering based solely on the information contained in the data. They usually depend on the representation of data (choice of features) and cannot be compared for different representations. *External* (or extrinsic) evaluation measures use some kind of external knowledge. This can be the existence of known classes (ground truth labelling) or the relevance judgements of human experts [68, 127].

Internal measures can give us information about how similar the generated clusters are inside (cohesion), how distinct from other clusters (separation), or both. Obviously, having only singleton clusters will lead to a situation where perfect homogeneity within clusters is achieved. Having only one cluster means that the dissimilarity with other clusters reaches a zero level. Both situations are thus optimal from one of the perspectives. However, since these two possibilities often do not match the reality, they are not desirable [244]; the number of clusters should be low [115] and some compromise needs to be found. The above mentioned clustering criterion functions can serve as internal measures of clustering quality. Another well known method is the silhouette method.

For external cluster evaluation, some kind of human judgement is often needed. The problem is that human judgements can be very subjective because everyone can have a different opinion about the topic. A correct answer to a clustering problem is also often not available. External evaluation looks at the closeness to a clustering provided by a human. Human experts thus usually create clusters that are compared to the clusters generated by an algorithm. Typical measures include purity, entropy, normalized mutual information, or F-measure.

The assignment of objects to clusters in the calculated clustering $U$ and the real (or expected) clustering $V$ can often be expressed in the form of a contingency table; see Figure 12.7. In the solution $U$, there exist $R$ clusters, $S$ clusters are defined in the $V$ clustering. The value on the $i^{th}$ row and $j^{th}$ column denotes the number of objects belonging to clusters $U_i$ and $V_j$ [267]. We can say that for these objects, the found and real assignment to a cluster agree. Based on the numbers from the contingency table, the values of some external cluster validity measures can be calculated.

|       | $V_1$    | $V_2$    | $\cdots$ | $V_S$    |
|-------|----------|----------|----------|----------|
| $U_1$ | $n_{11}$ | $n_{12}$ | $\cdots$ | $n_{1S}$ |
| $U_2$ | $n_{21}$ | $n_{22}$ | $\cdots$ | $n_{2S}$ |
| $\vdots$ | $\vdots$ | $\vdots$ | $\ddots$ | $\vdots$ |
| $U_R$ | $n_{R1}$ | $n_{R2}$ | $\cdots$ | $n_{RS}$ |

**Figure 12.7:** A contingency table characterizing an agreement between two clustering solutions, $U$ and $V$.

When evaluating clustering, compactness inside the clusters and separation of clusters are usually considered. Other factors, such as geometric or statistical properties of the data, the number of data objects, and dissimilarity or similarity measurements can be considered too [45]. Besides satisfying the homogeneity and separation criteria, clusters should also be complete. This means that objects from one class should be placed to one cluster. Another desirable property of clustering is to assign items not belonging to classes that are contained in clusters to a separate cluster. This cluster, containing 'other' or 'miscellaneous' elements, does not introduce disorder to existing clusters [13].

## 12.14.1  Metrics Based on Counting Pairs

One possibility of evaluating clusters is to calculate the statistics based on looking at how pairs of objects are placed to clusters in a clustering solution and a predefined partition [13]. Let there be $n$ objects $x_1, ..., x_n$ clustered to clusters $C_1$, ..., $C_k$. In addition, we have a defined partition (ground truth) $P_1$, ..., $P_l$. Looking at each pair $(x_i, x_j)$ from the data set, the following situations can occur:

■ SS: the objects belong to the same cluster of $C$ and to the same cluster of $P$.

■ SD: the objects belong to the same cluster of $C$ and to different clusters of $P$.

■ DS: the objects belong to different clusters of $C$ and to the same cluster of $P$.

■ DD: the objects belong to different clusters of $C$ and to different clusters of $P$.

$SS + SD + DS + DD$ is the number of pairs of data which equals to $n(n-1)/2$. $SS$ and $DD$ are good assignments, $SD$ and $DS$ are bad assignments.

The degree of similarity between a created clustering solution and the ground truth can be characterized by the following indices:

$$Rand\ Statistic = \frac{SS + DD}{SS + SD + DS + DD} \tag{12.37}$$

$$Jaccard\ Coefficient = \frac{SS}{SS + SD + DS} \tag{12.38}$$

$$Folkes\ and\ Mallows\ Index = \sqrt{\frac{SS}{SS + SD} \cdot \frac{SS}{SS + DS}} \tag{12.39}$$

High values of these indices indicate a good match between $C$ and $P$ [110].

## 12.14.2  Purity

*Purity* measures the fraction of a majority class in a cluster $C_i$:

$$Purity(C_i) = \max_{j \in L} \frac{l_{ij}}{|C_i|}, \qquad (12.40)$$

where $L$ is a set of classes, $l_{ij}$ is the number of documents from class $j$ in cluster $C_i$, and $|C_i|$ is the number of documents from cluster $C_i$. When a cluster contains instances from only one class, it is perfectly pure (clean) and the value of purity equals to 1. Purity of the entire clustering solution can be calculated as a weighted average of purities of all clusters:

$$Purity = \sum_{i=1}^{k} \frac{|C_i|}{n} Purity(C_i), \qquad (12.41)$$

where $k$ is the number of clusters and $n$ is the number of documents in the clustering solution.

## 12.14.3  Entropy

*Entropy* is an alternative measure expressing how many clusters contain documents from one class. Entropy, as proposed by Shannon, measures the average amount of information that is received in incoming messages; see section 14.3.2. The lowest entropy (value 0) is achieved when only one possible outcome exists. The highest value is in the case that all outcomes are equally likely. Entropy can thus be used for measuring the quality of clusters [256]. The best quality of a cluster is achieved when the cluster contains instances from one class. In this case, there are no surprises and the entropy is 0. The highest entropy {value $\log(c)$, where $c$ is the number of classes} and thus the worst quality of a cluster is obtained when there are instances of more than one class in the cluster and the numbers of documents from each class are equal. To calculate entropy, probabilities $p_{ij}$ that a document $j$ belongs to each of classes $j \in L$ must be calculated first:

$$p_{ij} = \frac{l_{ij}}{|C_i|}, \qquad (12.42)$$

where $l_{ij}$ is the number of documents from class $j$ in cluster $C_i$ and $|C_i|$ is the number of documents from cluster $C_i$.

Entropy of the cluster $i$ is then calculated as

$$Entropy(C_i) = \sum_{j \in L} p_{ij} \log p_{ij} \qquad (12.43)$$

To get the entropy of the entire clustering solution, we calculate the sum of the entropies of all clusters weighted according to their sizes:

$$Entropy = \sum_{i=1}^{k} \frac{|C_i|Entropy(C_i)}{n}, \qquad (12.44)$$

where $k$ is the number of clusters and $n$ is the number of instances to be clustered.

## 12.14.4 F-Measure

*F-measure* is often used in information retrieval and classification but its idea is suitable for evaluating clustering as well. It combines the values of the precision (the fraction of documents from a desired class that are placed to a cluster) and recall (the fraction of documents that should be put to a cluster and are really there). Both values are always calculated with respect to the given cluster and class. The precision for class $j$ in cluster $C_i$ can be calculated as

$$Precision(i, j) = \frac{l_{ij}}{|C_i|}, \qquad (12.45)$$

where $l_{ij}$ is the number of instances of class $j$ in cluster $C_i$ and $|C_i|$ is the total number of instances in cluster $C_i$. The recall for class $j$ in cluster $C_i$ is calculated as

$$Recall(i, j) = \frac{l_{ij}}{l_j}, \qquad (12.46)$$

where $l_{ij}$ is the number of instances of class $j$ in cluster $C_i$ and $l_j$ is the total number of documents of class $j$. The value of F-measure is typically calculated as the harmonic mean of the precision and recall:

$$F(i, j) = 2\frac{Precision(i, j)Recall(i, j)}{Precision(i, j) + Recall(i, j)} \qquad (12.47)$$

The overall F-measure for the entire clustering solution is the averaged F-measure for all classes distributed over all clusters [255]:

$$F = \sum_{j=1}^{l} \frac{l_j}{n} \max_{i=1..k} F(i, j) \qquad (12.48)$$

where $l$ is the number of classes, $l_j$ is the number of instances from the class $j$, $n$ is the number of all instances, and $k$ is the number of clusters. To achieve high values of the F-measure, both the precision and recall need to be high. Then the value of the F-measure will be close to 1.

## 12.14.5 Normalized Mutual Information

*Normalized Mutual Infromation* (NMI) belongs to the class of information theoretic measures. It measures how much information two clusterings, one found by an algorithm and the other created by a human, share. In other words, NMI tells us how much knowing about one of the clustering solutions reduces uncertainty about the other. Mutual information $I(U,V)$ of clusterings $U$ and $V$ is the difference between entropy of $U$ (the average amount of information needed to encode the labels of every object in $U$) and the entropy $H(U|V)$, which is the average amount of information needed to encode the labels of objects in $U$, knowing $V$:

$$I(U,V) = H(U) - H(U|V) \tag{12.49}$$

Alternatively, mutual information can be expressed as the sum of entropies of $U$ and $V$ minus the joint entropy of $U$ and $V$:

$$I(U,V) = H(U) + H(V) - H(U,V) \tag{12.50}$$

The entropy of $U$ can be calculated as follows:

$$H(U) = -\sum_{i=1}^{k} p_i log(p_i), \tag{12.51}$$

where $p_i$ is the probability of an object being in cluster $C_i$, i.e., $p_i = |C_i|/n$ ($|C_i|$ is the size of cluster $C_i$ and $n$ is the number of all objects).

$$H(U|V) = -\sum_{i=1}^{R}\sum_{i=1}^{S} \frac{n_{ij}}{n} log \frac{n_{ij}/n}{n_j/n}, \tag{12.52}$$

where $R$ is the number of clusters in $U$, $S$ is the number of clusters in $V$, $n_{ij}$ the number of objects placed to the same cluster in both $U$ and $V$, and $n_j$ is the number of objects in cluster $V_j$.

The joint entropy of $U$ and $V$ can be expressed as:

$$H(U,V) = -\sum_{i=1}^{R}\sum_{i=1}^{S} \frac{n_{ij}}{n} log \frac{n_{ij}}{n} \tag{12.53}$$

The mutual information can be expressed by the following formula:

$$I(U,V) = -\sum_{i=1}^{R}\sum_{i=1}^{S} \frac{n_{ij}}{n} log \frac{n_{ij}/n}{n_i n_j/n^2}, \tag{12.54}$$

where $n_i$ is the number of objects in cluster $U_j$ and the meaning of the other symbols is the same as in the previous formula [267, 12].

To be able to use mutual information for comparing different clustering solutions, a normalized version is often used. A few normalization schemes already exist. $NMI(U,V) = 2I(U,V)/(H(U)+H(V))$ [185] is probably the most popular normalization which scales the values of mutual information to the interval [0.1].

## 12.14.6 Silhouette

The *silhouette* method was proposed by Rousseeuw [231]. It looks at every single object and calculates how well is it located in its cluster and separated from the other clusters. To construct a silhouette, any distance metric can be used.

For every object $d_i$ from cluster $C_i$, the average distance $a_i$ (dissimilarity) of $d_i$ from all other objects in $C_i$, is calculated. The smaller the value, the better. A small value means that the objects in a cluster are homogeneous, closer to each other.

Subsequently, similar values are calculated for $d_i$ and all other clusters $C_j$, $j = 1...k$, $j \neq i$ (i.e., average distances of $c_i$ from objects in other clusters). The higher the value for a cluster $C_j$ the better it is because $d_i$ is different from the objects in that cluster. The minimal value $b_i$ from these average distances denotes the closest cluster (neighbor), i.e., the second best cluster choice for object $d_i$.

The silhouette for $d_i$ is calculated from $a_i$ and $b_i$ as follows:

$$s(i) = \begin{cases} 1 - \frac{a_i}{b_i} & \text{if } a_i < b_i \\ 0 & \text{if } a_i = b_i \\ \frac{b_i}{a_i} - 1 & \text{if } a_i > b_i \end{cases} = \frac{b_i - a_i}{max\{a_i, b_i\}} \quad (12.55)$$

From the formula, it is obvious that the value of $s(i)$ lies between $-1$ and $1$ (both values are included). When cluster $C_i$ contains only one object, the value of $a_i$ cannot be calculated so it is usually 0.

When the value of $s(i)$ is close to 1, the dissimilarity inside the cluster $a_i$ is much lower than the dissimilarity between clusters. This means that the object is well clustered (assigned to a correct cluster) because the objects in the second best cluster are much more dissimilar.

If the value of $s(i)$ close to $-1$, the object has most likely been placed to a wrong cluster; it would be better to have it in another cluster where the objects are more similar to it on average.

Value 0 indicates that the object is exactly between the best and the second best clusters and it could be put to either of them ($a_i$ and $b_i$ are equal).

When similarities need to be used instead of dissimilarities, the formulae for calculating the value of $s(i)$ need to be modified:

$$s(i) = \begin{cases} 1 - \frac{b_i'}{a_i'} & \text{if } a_i' > b_i' \\ 0 & \text{if } a_i' = b_i' \\ \frac{a_i'}{b_i'} - 1 & \text{if } a_i' < b_i' \end{cases} \quad (12.56)$$

The silhouette method is based on a principle similar to the group average linkage methods so it works well for roughly spherical clusters.

A great advantage of the silhouette method is the ability to visualize the correctness of assignment of objects in clusters. A typical graphical representation is a plot where every object is depicted by one horizontal line with the size proportional to the value of $s(i)$. If the value of $s(i)$ is negative, the line is drawn in the opposite direction to the lines representing objects with positive values of $s(i)$. All lines are arranged one above the other, beginning at the same position. The objects in one cluster are in one group where the objects are sorted according to the value of $s(i)$ with the highest value first.

Intuitively, higher values of $s(i)$ are preferred to lower values. Thus, cluster 1 in Figure 12.8 is better than cluster 2 because its silhouette is wider. It is also desirable that there are no objects with low values of $s(i)$. Examples of this are the last two objects in cluster 3 where the last one should even be put to a different cluster as it is more similar to a cluster other than cluster 3.

The silhouette method can be also used to find a correct number of clusters. The silhouettes for different clustering solutions can be analyzed and the solution with the best cluster silhouettes is selected.

If the number of clusters to be created is too high, some naturally occurring clusters will be split into more clusters (in a situation where the natural clusters are very homogeneous and far from the others). For example, objects in a natural cluster A will be placed to clusters B and C. These clusters will still be quite

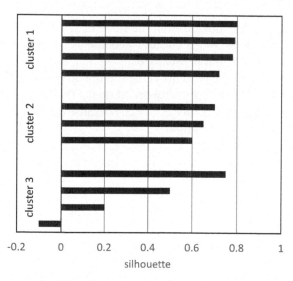

**Figure 12.8:** An example of the silhouette values for a clustering solution of several objects into three clusters.

homogeneous but very similar to each other. Thus, the values of $s(i)$ will be rather small compared to the situation when we have only cluster A.

When the number of clusters is too low, some of the clusters that should be separated will be merged. Now, the internal similarity will be higher, which will again lead to lower values of $s(i)$.

We can also calculate an average silhouette width for each cluster. This is simply the arithmetic average of the values of $s(i)$ for all objects in cluster $C_j$:

$$s(C_j) = \frac{1}{|C_j|} \sum_{x \in C_i} s(x) \tag{12.57}$$

Clusters with high values of average silhouette width are more obvious than those with lower values. We can even calculate the average silhouette width for all objects in the entire data set:

$$s(k) = \frac{1}{k} \sum_{r=1}^{k} s(C_r) \tag{12.58}$$

The clustering solution with the highest value [8] will be preferred.

## 12.14.7 Evaluation Based on Expert Opinion

When no prior knowledge about the expected outcomes of clustering is available, an expert can inspect the results and decide about their usefulness. This kind of validation carried out by human experts is suitable under many circumstances. It may reveal new insights into the data, but is generally very expensive and demanding. The results, which are subjectively influenced are also not comparable.

The objects to be clustered are characterized by some attributes and their values. Each object can be thus seen as a multidimensional vector or a point in a multidimensional space where the number of dimensions is equal to the number of attributes. Intuitively, the clusters can be defined as sub-spaces containing the objects which are clearly separated from the other sub-spaces. This identification can be thus done visually by a human expert. However, the problem is when the number of dimensions is greater than three. There are methods that are able to project objects from a multidimensional space to a two- or three-dimensional space. This enables the users to examine visually whether the created clusters correspond to the clusters that they can see. Such an approach is based on the principle that groups of objects that are separated in 2D or 3D are also separated in a space with many dimensions [122]. This kind of evaluation is, however, difficult to use for data with a very high number of dimensions, which is the case of textual document collections.

The experts might also examine the objects in individual clusters and consider the extent to which such an arrangement contributes to the objective of a given task. In the case that many objects are clustered, such a manual process would be very difficult and thus unfeasible. In that case it is possible to not consider all of the documents but examine only some of them. There are several approaches to choose the representative documents [99]. A user might be presented with a document which is:

- an average document – the document lies most closely to the remaining documents in the cluster, i.e., it is located near the centroid of the cluster;

- the least typical element – the document is close the border of the cluster, closer to the remaining clusters; it is most similar to documents in the other clusters;

- the most typical – the document is located near the border of the cluster, far from the other clusters; it is the most different from all of the documents in the collection and thus the most specific.

The documents found by using the last two approaches might be difficult to interpret, especially in the case that there are many clusters in the clustering solution. Therefore, choosing a document that is located in the center of a cluster seems to be appropriate.

Sometimes, one document may not be informative enough. Therefore, more documents selected using the same criterion might be used. In that situation, the documents can be ranked according to the given criterion and a few documents which lie at the top of such a list of ranked documents can be presented.

## 12.15   Cluster Labeling

Sometimes, not only are the labels for the data not assigned to the documents but the labels may not even exist. Consider a company selling some products or providing some services. On some web pages, the customers have the facility to talk about them, evaluate them, or simply send complaints electronically, everything freely written using a natural language. Reading thousands of such messages by a human expert and identifying the major topics that are mentioned would not be feasible. Instead of that, a computer might try to find groups of documents that are somehow similar and distinguished from the other groups. These groups will be more or less related to individual aspects of the product or service. Subsequently these groups might be further processed so that typical representatives of the groups or keywords characterizing the groups can be retrieved and used to, for example, improve the business quality. Such keywords are often referred to as *cluster labels*. It is generally good to characterize the clusters using a small number of variables [115]

According to Ferraro and Wanner [88], two main strategies of document cluster labeling can be identified. In *internal cluster labeling*, which is usually quite a simple approach, the label of the cluster is based solely on the content of the cluster. The title of a representative document or a list of words that appear with a sufficiently high frequency might be used. In *differential cluster labeling*, the label is determined by contrasting the cluster with other clusters. When a candidate label depends on a cluster more than on the other clusters, it is considered a good label for that cluster. Several statistical measures, such as mutual information, information gain, or chi squared might be applied.

## 12.16   A Few Examples

The following example shows how to cluster a document set, determine a suitable number of clusters, evaluate alternative clusterings, and characterize the found clusters.

In the first example, only eight documents that were manually created are processed. Here, a reader clearly sees the content of the documents and makes decisions about assigning them to clusters. The documents, most naturally, form tree clusters:

■ One is about *clustering*:

 ■ Document 1: *Clustering is an unsupervised machine learning method and assigns data to groups.*

 ■ Document 2: *Clustering finds groups in data in an unsupervised way.*

■ The other is related to *classification*:

 ■ Document 3: *Classification is a supervised machine learning method and assigns a class to an object.*

 ■ Document 4: *A class is determined in a supervised classification.*

■ The last one mentions *statistics* (the last two documents are more similar than the first two because they both mention analysis and summarization):

 ■ Document 5: *Statistical analysis can be applied to large data sets.*

 ■ Document 6: *Statistics deals with large data sets analysis.*

 ■ Document 7: *Statisticians analyze and summarize large data sets.*

 ■ Document 8: *It is fun to analyze and summarize data.*

To be able to cluster the data, a conversion to a numerical document-term matrix is performed using the tm package.

The data is then clustered using several algorithms. Some of them work with a data matrix representing the objects to be clustered, some others accept a dissimilarity matrix containing pairwise dissimilarities of all pairs of objects. The dissimilarity matrix can be calculated using the dist() function from the stats package or with the daisy() function from the cluster package.

The kmeans() function from package stats is used to cluster the data in a matrix using the k-means algorithm. The function requires a numeric matrix of data (or an object that can be coerced to such a matrix), the number of clusters to be found, or a set of initial cluster centres. If the initial centers are not provided, a random set of rows in the data matrix is chosen as the initial centres. We might also supply the maximum number of iterations of the algorithm.

The function returns an object of class kmeans. Its components include cluster – a vector of integers indicating the cluster to which each point belongs, centers – a matrix of cluster centres, or size – the number of objects in each cluster.

In our example, values 2, 3, and 4 are successively supplied as the parameters determining the number of clusters. For each clustering solution, assignment of the documents to clusters is displayed by printing the contents of the cluster element of the clustering list. Then, the silhouette() function from package cluster is used to calculate the values of the silhouette for each document. The first parameter of the function contains an object of an appropriate class, an integer vector with k different integer cluster codes or a list with such a clustering component by default. The other parameters include a dissimilarity matrix (dist), that can be in its symmetric form (dmatrix). The function returns a silhouette class object, which is an n x 3 matrix with attributes. Each row (document) contains the cluster to which it belongs, the neighbour cluster (the cluster not containing the document, for which the average dissimilarity is minimal), and the silhouette width (the column names are cluster, neighbor, and sil_width). The values of silhouette can be sorted so they can be displayed in the usual way – objects from one cluster together, in a descending order.

Function sortSilhouette() orders the rows of a silhouette object increasingly by cluster and with decreasing silhouette width. To visualize the silhouette, the standard plot() function can be used. We can also use the package factoextra that provides functions that can be used to extract and visualize the results of multivariate data analyses. One of them, fviz_silhouette(), visualizes silhouette information from clustering.

We can see that the solution with three clusters has the highest value of silhouette, significantly higher than for the solution with two clusters. The silhouette for the solution with three clusters also looks much better that the solution with two clusters. Four clusters seem to be only slightly worse than three clusters as the last four documents can be considered one or two clusters.

As the k-means method used in this example depends on some random initialization, a reader can sometimes obtain different results (typically, the documents will be differently distributed in the generated clusters).

```r
library(tm)
library(ggplot2)
library(factoextra)

# converting texts to a structured representation
texts <- c("Clustering is an unsupervised machine
            learning method and assigns data to groups",
           "Clustering finds groups in data in an
               unsupervised way",
           "Classification is a supervised machine learning
            method and assigns a class to an object",
           "A class is determined in a supervised
            classification",
           "Statistical analysis can be applied to large
            data sets",
           "Statistics deals with large data sets analysis",
           "Statisticians analyze and
            summarize large data sets",
           "It is fun to analyze and summarize data.")
documents <- data.frame(doc_id=1:8,
                        text=texts,
                        stringsAsFactors = FALSE)
corpus <- Corpus(DataframeSource(documents))
corpus <- tm_map(corpus, removePunctuation)
corpus <- tm_map(corpus, content_transformer(tolower))
dtm <- DocumentTermMatrix(corpus,
           control = list(
           wordLengths = c(1, Inf),
           weighting=function (x) weightSMART(x,spec="ntn")
           )
)

# a matrix to be used by kmeans()
mat <- as.matrix(dtm)

# a distance matrix to be used for silhouette calculation
distances <- dist(mat)

# an object for storing silhouette values for
# different partitionings
silhouette <- matrix(0, nrow=4, ncol=8)
# trying three different partitionings
for (nclust in 2:4) {
```

```r
print(paste("Number of clusters:", nclust))
clustering <- kmeans(mat, nclust)

# looking at assignment of document to clusters
print("Cluster assignment:")
print(clustering$cluster)

# looking at most important features characterizing
# a cluster
for (c in 1:nclust) {
  print(paste(paste("Cluster", c), ":"))
  print(sort(clustering$centers[c,], decreasing=TRUE)[1:5])
}

# computing sihlouette information
sil <- silhouette(clustering$cluster, dist=distances)

# storing the values of silhouette for a clustering result
silhouette[nclust, ] <- as.vector(sortSilhouette(sil)[,3])

# plotting a silhouette graph
plot(fviz_silhouette(sil, label=TRUE, print.summary=TRUE))
}

# printing silhouette information for different partitionings
for (nclust in 2:4) {
  print(paste("Silhouette for cluster ", nclust))
  print(round(silhouette[nclust, ], 3))
  print(paste("Mean:", round(mean(silhouette[nclust, ]), 3)))
}
```

The output obtained from the executed statements gives us information on how the documents were assigned to clusters (each cluster is labeled with a number, the numbers start from 1), what features characterize them well, silhouette information, and plots with visualized silhouette information (see Figure 12.9):

```
[1] "Number of clusters: 2"
[1] "Cluster assignment:"
1 2 3 4 5 6 7 8
1 1 1 1 2 2 2 2
[1] "Cluster 1 :"
        a          in          an   assigns clustering
```

```
 2.000000   1.500000   1.061278   1.000000   1.000000
[1] "Cluster 2 :"
    large        sets  analysis   analyze summarize
 1.061278   1.061278   1.000000   1.000000   1.000000
   cluster size ave.sil.width
1       1    4          0.08
2       2    4          0.14


[1] "Number of clusters: 3"
[1] "Cluster assignment:"
1 2 3 4 5 6 7 8
2 2 1 1 3 3 3 3
[1] "Cluster 1 :"
              a       class classification   supervised
            4.0         2.0            2.0          2.0
        object
            1.5
[1] "Cluster 2 :"
  clustering       groups unsupervised          in
         2.0          2.0          2.0         2.0
       finds
         1.5
[1] "Cluster 3 :"
    large        sets  analysis   analyze summarize
 1.061278   1.061278   1.000000   1.000000   1.000000
   cluster size ave.sil.width
1       1    2          0.24
2       2    2          0.13
3       3    4          0.11


[1] "Number of clusters: 4"
[1] "Cluster assignment:"
1 2 3 4 5 6 7 8
2 2 1 1 4 4 3 3
[1] "Cluster 1 :"
              a       class classification   supervised
            4.0         2.0            2.0          2.0
        object
            1.5
[1] "Cluster 2 :"
  clustering       groups unsupervised          in
         2.0          2.0          2.0         2.0
       finds
```

```
          1.5

[1] "Cluster 3 :"
        analyze     summarize statisticians            fun
           2.0           2.0            1.5            1.5
            it
           1.5
[1] "Cluster 4 :"
   analysis       applied           be    can statistical
        2.0           1.5          1.5    1.5         1.5
   cluster size ave.sil.width
1        1    2          0.22
2        2    2          0.08
3        3    2          0.26
4        4    2         -0.03

[1] "Silhouette for cluster  2"
[1] 0.162 0.085 0.040 0.034 0.162 0.147 0.143 0.117
[1] "Mean: 0.111"
[1] "Silhouette for cluster  3"
[1] 0.245 0.243 0.189 0.065 0.130 0.123 0.117 0.089
[1] "Mean: 0.150"
[1] "Silhouette for cluster  4"
[1]  0.243  0.206  0.151  0.014  0.293  0.217  0.010 -0.062
[1] "Mean: 0.134"
```

A partitioning algorithm that clusters data around medoids instead of centroids is also available in R. It is implemented in the pam() (Partitioning Around Medoids) function in the cluster package. The most important parameters include the data to be clustered (a data matrix, data frame, dissimilarity matrix, or an object), the number of clusters, a metric to calculate similarity (Euclidean or Manhattan), and a parameter defining initial medoids so they are not automatically found in the initial phase of the method.

Function agnes() implements agglomerative nesting hierarchical clustering. The first parameter contains the data to be clustered. By default, it is a dissimilarity matrix. The parameter can also contain a matrix or data frame where each row represents one observation and each column a variable.

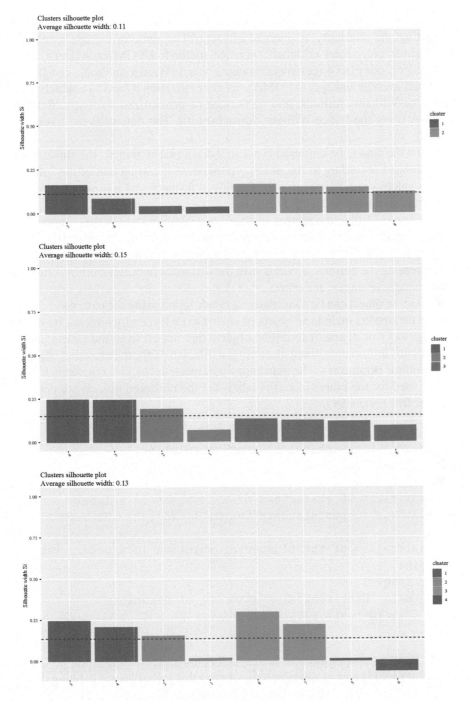

**Figure 12.9:** Visualized silhouette information for three different clusterings of the same data set.

The function creates a dendrogram using one of the seven options for calculating the similarity of merged clusters. The options include average (unweighted pair-group arithmetic average method, UPGMA), single (single linkage), complete (complete linkage), ward (Ward's method), weighted (weighted average linkage, WPGMA), its generalization flexible which uses the Lance-Williams formula, and gaverage (a generalized average, flexible UPGMA using the Lance-Williams formula) [20]. The default is the average option.

When the data to be clustered is not in a dissimilarity matrix, the similarities are calculated according to the metric parameter. It can be one of the following two values – euclidean for the Euclidean distance and manhattan for the Manhattan distance.

The agnes() function returns an object of class agnes with the components holding the information about the clustering results, such as the order of objects in a graphical plot (so the branches of the dendrogram do not cross), information about merging the clusters, a vector with the distances between merging clusters at the successive stages, and others.

An agnes object can be converted to a dendrogram using the pltree() function. Its parameters include an object of class twins (typically returned by functions agnes() or diana() returning objects that inherit from the twins class) that contains information about clustering, a title, labels for the clustered objects, and graphical parameters. The mandatory parameter is the first one, default values are used for the others (e.g., the labels for the clustered objects are derived from the first parameter).

```
# finding a hierarchical clustering solution
hc <- agnes(distances)

# taking first 25 characters of a document as a label
texts.labels <- paste(substr(texts, 1, 25), "...", sep="")

# visualization of the clustering solution in a dendrogram
pltree(hc,
        hang=-1,
        main="Hierarchical clustering of documents",
        labels=texts.labels[hc$order])
```

The hang parameter describe how much below the rest of the plot the labels of the plot should hang. A negative value means that the labels are hanging down from 0. The labels parameter assigns labels to the clustered objects.

A dendrogram visualizing agglomerative clustering of the eight documents can be found in Figure 12.10.

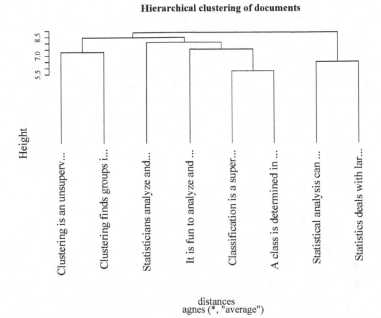

Figure 12.10: Visualized dendrogram for the agnes() method.

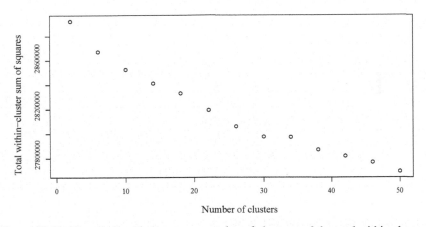

Figure 12.11: The relationship between a number of clusters and the total within cluster sum of square errors when clustering 50,000 customer reviews from booking.com. An elbow is obvious for the number of clusters equal to 30.

A divisive hierarchical algorithm is implemented by the diana() function (DIvisive ANAlysis). This algorithm starts with all items in one cluster. The clusters are subsequently divided until each cluster contains only a single item. At each stage, the cluster with the largest diameter (the largest dissimilarity between any two of its observations) is selected.

The first parameter contains the data to be clustered. By default, it is again a dissimilarity matrix: a matrix, or data frame where each row represents one observation and each column a variable. The `metric` parameter, which can be one of the following two values – `euclidean` for the Euclidean distance and `manhattan` for the Manhattan distance is used to calculate dissimilarities.

The following example tries to find aspects of the hotel accommodation service, which is in fact an automatic detection of semantic content of groups of textual documents. The documents here are customer reviews written by people who really stayed in a hotel and have had real experience with it. The data comes from the famous server *booking.com* and a detailed description might be found in [295]. In our experiment, 50,000 reviews written in Englsh were used in the analysis.

We can assume that similar documents share a common topic, in this case, an aspect of the service. To find groups of similar documents, clustering with the k-means algorithm was used. The elbow method analyzing the dependence between a number of clusters and the within cluster sum of square errors was used to find a right number of clusters needed by the k-means algorithm (30 clusters were chosen). Differential cluster labeling using the chi squared method (see Section 14.3.1 for details) was used to find representative words characterizing each cluster.

```
library(tm)

# reading the reviews and creating a document-term matrix
texts <- readLines("reviews-booking.txt")
documents <- data.frame(doc_id=1:length(texts),
                        text=texts,
                        stringsAsFactors=FALSE)
corpus <- Corpus(DataframeSource(documents))
corpus <- tm_map(corpus, removePunctuation)
corpus <- tm_map(corpus, content_transformer(tolower))

dtm <- DocumentTermMatrix(corpus,
          control=list(
              wordLengths=c(1, Inf),
              removeNumbers=TRUE,
              stopwords=TRUE,
              bounds=list(global=c(11, Inf)),
              weighting=function (x)
                      weightSMART(x, spec="ntn")
```

```
            )
)
mat <- as.matrix(dtm)

# calculating the within cluster sum of square errors for
# different numbers of clusters
withinss <- c()
for (k in seq(2, 50, 4)) {
    clustering <- kmeans(mat, k, iter.max=100)
    withinss[k] <- clustering$tot.withinss
}


# visualizing the relationship between a number of clusters
# and the within cluster sum of square errors to find an
# elbow denoting a suitable number of clusters
plot(withinss,
     xlab="Number of clusters",
     ylab="Total within-cluster sum of squares")

# 30 seems to be a good number of clusters
nclust <- 30
clustering <- kmeans(mat, nclust)

# calculating the values needed for Chi squared
# values estimation
D<-C<-B<-A<-matrix(nrow=nclust, ncol=dim(dtm)[2])
rownames(D)<-rownames(C)<-rownames(B)<-rownames(A)<-
            c(1:nclust)
colnames(D)<-colnames(C)<-colnames(B)<-colnames(A)<-
            colnames(dtm)

for (c in c(1:nclust)) {
   for (w in c(1:dim(dtm)[2])) {
      A[c,w] <- sum(dtm[clustering$cluster == c, w] != 0)
      B[c,w] <- sum(dtm[clustering$cluster != c, w] != 0)
      C[c,w] <- sum(dtm[clustering$cluster == c, w] == 0)
      D[c,w] <- sum(dtm[clustering$cluster != c, w] == 0)
   }
}

# calculating the values of the Chi square metric
# (not in one formula to prevent integer overflow due
# to very high numbers)
```

```
chi <- dim(dtm)[1]*((A*D)-(C*B))^2
chi <- chi/(A+C)
chi <- chi/(B+D)
chi <- chi/(A+B)
chi <- chi/(C+D)

# calculating the average tf-idf of terms in the collection
avg_frequencies <- colMeans(mat)

# calculating the average tf-idf of terms in each class
tab <- rowsum(mat, clustering$cluster)
numbers_of_documents_in_clusters <- clustering$size
tab <- tab/numbers_of_documents_in_clusters

# printing 10 most important attributes for each class
# (only attributes with the average tf-idf value for a class
# higher than the average for all classes)
print("The values of Chi square for attributes and classes:")
for (c in c(1:nclust)) {
    print(paste("Class: ", c), sep="")
    print((sort((chi[c, tab[c, ] > avg_frequencies]),
               decreasing=TRUE))[1:10])
}

# printing 5 documents closest to cluster centroids
for (c in c(1:nclust)) {
    print(paste("Cluster ", c, sep=""))
    distances <- apply(mat[c==clustering$cluster, ],
                       1,
                       function(x)
                           sum((x - clustering$centers[c, ])^2)
                       )
    print(texts[as.numeric(names(sort(distances)[1:5]))])
}
```

The output of the statements above contains the list of the ten most discriminative words that are contained in documents from each cluster.

For some of the clusters, one can quite easily assign one clear label describing the main content well. For example, cluster 2 is about *television*, cluster 5 about *indoor pool and gym*, cluster 8 about *airport shuttle*. Other clearly identified aspects of the accommodation service include *parking, credit cards, furniture, expensive bar*, *beds*, *staff*, etc. Examples of a few clusters with good labels follow (in fact, for most of the 30 clusters, a topic could be easily determined):

```
...
[1] "Class: 2"
   channels          tv         bbc        news         cnn
36592.5493 11672.0444  2430.7969  2413.2943  1566.8983
   english     channel    language     digital  television
 1221.4890    890.0239    496.8955    402.3057    267.6288
...
[1] "Class: 5"
       pool    swimming      indoor        swim         gym
39290.3494 13360.0888  1395.6040    693.1596    515.0593
    outdoor         spa       sauna     swiming     jacuzzi
   484.8979    413.5477    367.6770    348.9054    334.2497
...
[1] "Class: 8"
    airport     shuttle        taxi         bus    transfer
40243.6288  5033.6403  1259.5266    960.2551    758.1955
     flight       close      fromto      driver     gatwick
   727.6716    543.1865    486.8610    377.9844    346.0994
...
[1] "Class: 9"
      dirty      carpet     carpets      sheets      stains
41349.3561  1481.7790    585.0428    446.2469    361.6065
     filthy    bathroom       blood     cobwebs   wallpaper
   337.5244    277.6227    275.5182    271.5584    248.0691
...
[1] "Class: 25"
    parking         car        park        free      garage
29786.0432  7737.2175  4017.3253    805.5700    762.7539
      space      spaces      secure        cars   difficult
   710.4421    710.2466    579.6436    569.6599    335.8032
...
```

Of course, the results are not perfect. Therefore, there were some clusters for which one clearly defined topic could not be easily determined as shown below. Cluster 11 in this example, was quite big compared to the others, so further analysis of it could bring additional insights to the data.

```
. . .
[1] "Class: 3"
 friendly      staff    helpful       good   location
 7854.677   7240.694   4325.775   4049.060   3163.018
excellent      clean      value       nice      great
 2919.974   2852.039   1783.016   1656.075   1529.375
. . .
[1] "Class: 11"
   improved friendliness     cleaness   cleaniness
  18.152680    14.861314    12.864497    11.573488
   updating  convenience    closeness     victorian
  11.235698     8.744312     7.423451     7.086854
   makeover     exciting
   7.042680     6.839037
. . .
[1] "Class: 13"
      one       asked        get       said        told
2316.2821   1819.5886  1637.1390  1294.0469  1276.1080
    night     another       even        got         day
1253.5418   1252.8707  1100.1629  1045.5220   985.9492
. . .
```

To get examples of whole documents from the clusters, the most typical were studied (most typical documents look close to the cluster centroids). Examples of such documents for several clusters can be found below.

```
. . .
[1] "Cluster 2"
[1] "Could have more TV channels"
[2] "Very good location..Very clean..friendly staff..English
channels on TV"
[3] "No English channels on the TV."
[4] "TV channels are a bit fuzzy"
[5] "Just only few TV channels"
. . .
[1] "Cluster 5"
[1] "There is no pool"
[2] "the pool"
[3] "a pool"
[4] "That there was no pool"
[5] "absense of pool"
. . .
```

```
[1] "Cluster 8"
[1] "distence from airport"
[2] "location to shannon airport."
[3] "It was close to the airport!"
[4] "very close to airport"
[5] "Close to Weeze Airport"

[1] "Cluster 9"
[1] "Dirty"
[2] "Toilet/Bathroom very dirty."
[3] "Dirty hotel"
[4] "Room was too small and dirty."
[5] "The hotel was very dirty& Room was very small"
...
[1] "Cluster 25"
[1] "Convient location,easy parking."
[2] "No Parking!"
[3] "no parking in the hotel"
[4] "Very good postion.easy parking"
[5] "The parking and location."
...
```

# Chapter 13

# Word Embeddings

## 13.1 Introduction

The main problem with the bag-of-words model is that it does not capture relations between words. In the bag-of-words model, each word or other feature of a text is represented by one dimension in a multidimensional space for representing the documents (we talk about *one-hot representation* [101]). Each such dimension is independent of the others because it is represented by only one value which does not enable sharing some information across features. It is, therefore, not possible to say that, for example, the word *football* is more similar to the word *soccer* than to the word *ballet*.

On the other hand, when there are relatively few features and there are no correlations between them (or we do not want to take the correlations into consideration), one-hot representations can be more beneficial than alternative representations which somehow represent the correlations and reduce the sparseness typical of the bag-of-words model [101].

To solve the problem where there is no similarity among words in the bag-of-words model, it would be possible to add other information to the existing words in a text, to better capture their context. Instead of keeping information, for example, *the word is A*, it is possible to store that *the word is A, it's POS tag is T, the previous word is B, and the following word is C*. This, however, increases the number of dimensions in the input space and requires the combination of possible feature components to be carefully selected.

Some of the modern representations of texts are based on the idea that similar words should have similar properties. This, however, cannot be achieved in the classic bag-of-words model. Thus, more than one dimension is needed to represent each word or feature. In this approach, each word is mapped to a continuous

I visited *Brno* last year.
He travelled from *Brno* to New York.
*Brno* has almost half a million inhabitants.
There are twenty nine districts in *Brno*.

**Figure 13.1:** Word *Brno* appearing in a few sentences. A reader can guess that Brno is a city even without ever hearing about it. This is because the word appears in the same or similar contexts as the names of cities the reader knows.

multidimensional space which typically has a few hundred dimensions, which is much less than the number of dimensions in the bag-of-words model. The words are then embedded in this continuous vector space. We therefore talk about *word embeddings*.

Since the *word vectors* are usually found in an unsupervised training process, the dimensions are difficult to interpret. The vector can be understood as a vector of contextual features, because it is constructed based on the context in which the word appears. Anyway, similar words in this space are expected to be close to each other.

The basic idea of these alternative models is that words appearing in similar contexts usually have similar meaning. For an explanation, look at Figure 13.1. Here, a few sentences contain the not very well-known word, *Brno*. Despite this, a reader understands that it is very likely the name of a city. This is because similar sentences contain names of cities which the reader knows. These sentences typically talk about the number of inhabitants, districts, travelling directions, etc. All these aspects need to be somehow contained in a representation of a word; hence, so it is obvious that one dimension in a bag-of-words model is definitely not enough.

Similar words share some information; therefore, the values of their vector elements should be similar. They are thus located close to each other in the multidimensional space. Besides capturing the similarity between the words, models based on this principle also bring some additional advantages or possibilities. Because words are represented by vectors, vector operations can be performed with them, which can provide very surprising answers.

If a relationship exists between words A and B, and there is a similar relationship between words C and D, the offset of A and B, and C and D is also similar; see Figure 13.2. The relationship can be semantic, like the gender relation, or city–country relation as in Figure 13.2, but also syntactic like singular–plural, present tense–past tense, or base–comparative–superlative forms of an adjective [195].

Probably the most famous example presented in [195] says that

$$vector("king") - vector("man") + vector("woman")$$

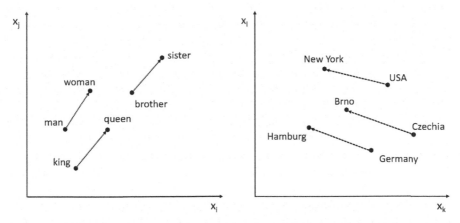

**Figure 13.2:** Examples of regularities in the word embeddings space.

will be very close to *vector("queen")* which captures the male/female relationship. The capital-of relation can be demonstrated with an example like

$$vector("paris") - vector("france") + vector("poland") \approx vector("warsaw")$$

and a pluralisation relation by

$$vector("cars") - vector("car") + vector("apple") \approx vector("apples")[269].$$

Using word vectors can also significantly help in calculating true document similarities. When two sentences have a very similar meaning but use synonyms instead of the same words classic similarity measures will say that the sentences are not similar. When stop words are not taken into consideration, the sentences *The queen visited the capital of US.* and *Elizabeth came to Washington in the USA.* share no words. They, however, describe the same situation. In the space of word vectors, word pairs *Elisabeth* and *queen* and *USA* and *US* will be very close. This is the idea behind the *Word Movers Distance* (WMD) measure, which quantifies the dissimilarity between two documents as the minimum amount of distance that the embedded words of one document need to travel to reach the embedded words of another document, thus taking similarity of word vectors into consideration [157].

## 13.2  Determining the Context and Word Similarity

Finding suitable values for the elements of the vectors representing words is based on the hypothesis stating that words in similar contexts have similar meanings [165]. A key question is, what is the context of a word. A context can be defined as a set of words that are, in some way, in a defined environment of the word. It might be the same document, sentence, or a piece of a text of some length (for example, five words to the left and five words to the right).

The simplest form of finding words with similar contexts is looking at the document-term matrix. If the matrix has $n$ rows ($n$ is the number of documents) and $m$ columns ($m$ is the number of terms), words with similar contexts have similar columns. In this case, the context are entire documents.

If the context is not the entire document, it is more common to use a *word co-occurrence matrix* instead of a document-term matrix. A word co-occurrence matrix is a square matrix where the rows and columns are labeled by the words. The elements of the matrix contain the frequencies with which the words co-occur.

A disadvantage of using raw frequencies in a term-term matrix is that some words with high frequencies (often stop words) tend to co-occur with many other words but are not very discriminative. Instead, it is better to somehow normalize the frequencies so that real relations between words become obvious. One of the more commonly used measures in this context is *pointwise mutual information* (PMI); see Section 14.3.2. It measures whether two words occur more often together than if they were independent. The values of PMI range from $-\infty$ to $+\infty$. Negative values are replaced by zeros because it can be problematical to interpret what it means that words co-occur less than by chance. The measure is biased towards low frequency words, so raising the probabilities to a number (e.g., 0.75) or adding one (Laplace smoothing) to all frequencies is often used [138].

Both document-term matrices and word co-occurrence matrices usually have many rows and columns (tens of thousands) and their vectors are sparse (many zeros). It is desirable to use representations where the vectors have different properties. They should be shorter so that using them in a machine learning task is easier (less parameters need to be learned) and most of the values are not zeros (i.e., the vectors are dense).

Popular approaches leading to generating dense vectors include models using global matrix factorization like *Latent Semantic Analysis* (LSA) or *Latent Dirichlet Allocation* (LDA) and models learned by neural networks using a small context window (where *word2vec* is probably the most popular). In many situations, the models based on neural networks perform better than LSA in the preservation of linear regularities among words (important for the word analogy task, for example) and are computationally more efficient than LDA. On the other hand, global methods like LSA can efficiently leverage statistical information unlike local methods like *word2vec* [195, 214].

## 13.3 Context Windows

Short pieces of text in a neighbourhood of a word are quite popular to define the context. Three context types can be distinguished according to what is being considered [101]:

■ Continuous bag-of-words – the context is represented by words from a window where no order or positions are considered (e.g., the context of X is A, B, C, and D).

■ Positional – the relative positions of other words are considered (e.g., the context of X is A two positions to the left, C one position to the left, D one position to the right, and C two positions to the right from X).

■ Dependency – syntactic connections to the word are considered (e.g., A is subject of X, B is object of X).

When the context is determined by a window surrounding a word, the windows size has an impact on the learned embeddings. When smaller context windows are used (1–3 words), more syntactic and functional similarities are found. Longer windows (4–10 words) discover more semantic similarities [138].

The content of the window can also be modified before training. The words can have different weights assigned according to the distance to the word for which they define the context (this can be achieved through sampling). Words of high frequency can be removed before training. Here, two cases can be distinguished – the words are removed before defining their contexts (which means that the context window enlarges, words further from the predicted word will be included in the context), or after creating the windows (the context window has the same size). Rare words can also be removed; this, however, has little impact on the quality of learned embeddings [166].

## 13.4 Computing Word Embeddings

Supervised methods require annotated data for a specific task; for example, texts with part of speech tags for the part of speech tagging task. Then, the embeddings are trained towards the given goal and can capture information that is relevant for the task. The created embeddings can be successfully used for a related task (like syntactic parsing) for which there is not enough labeled data. The embeddings learned in one task can be also transferred to another task and improved with some labeled data from the other task [101].

Unsupervised methods for word embeddings training do not require annotated data. Their only goal is to compute embeddings. The embeddings are usually learned in the task of predicting a word given its context, or deciding whether, given examples of real and randomly created word-context pairs, a word can belong to a context [101]. Large corpora that are available can be used for training instead of annotated texts. The learned embeddings capture general syntactic and semantic relationships and can be applied in a wide variety of tasks.

There are a few famous methods that can be used to compute word embeddings:

■ **Neural language models.** Dense vector representations (embeddings) can be learned in a neural network (usually the first layer) that is provided with one-hot vectors [101]. The first works producing word embeddings used neural networks for language modeling. In a neural probabilistic language model [21], $k$ words from a text were taken as an input and the probabilities of the following word were calculated using softmax; see Figure 13.4. Collobert et al. [57] used a window around a word instead of several preceding words to predict a correct word. They also replaced probability calculations with the assignation of higher scores to correct words and lower scores to incorrect words. Vectors representing the words are concatenated and processed by a neural network with one hidden layer.

■ **Word2vec.** *Word2vec* is a family of methods proposed by Tomáš Mikolov that drew the NLP community to neural language models. The prediction in *word2vec* has two forms – predicting the word based on its context (the Continuous bag-of-words or CBOW approach), and predicting the context for a word (the Skip-gram approach); see Figure 13.4. The input and output in both cases are one-hot encoded vectors. *Word2vec* tried to eliminate the problems of the previous neural language models with computational complexity. In the training phase, a neural network uses a linear activation function instead of the sigmoid function, which is typical for a multilayer perceptron, and the logarithm of the probability of predicting the word or its context context is maximized [194].

As Levy and Goldberg [165] found, learning word vectors using the skip-gram model with negative sampling (artificially created pieces of text that are added to the training set should be distinguished from real texts) are an implicit factorization of a positive pointwise mutual information matrix of word-context pairs.

■ **GloVe.** *GloVe* is a model developed at Stanford University. Unlike *word2vec*, which learns word vectors directly, *GloVe* uses information about global co-occurrences of words. It uses word vectors in the task of predicting the probability with which two words co-occur. More precisely, the ratios of word co-occurrence probabilities are predicted as the relationship between two words is examined by studying the ratio of co-occurrences with other, so called, probe words. The probabilities can be

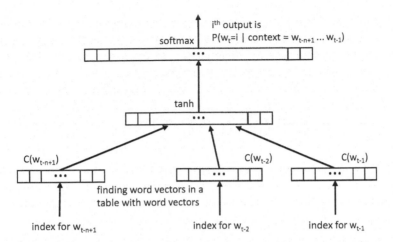

**Figure 13.3:** The architecture of a probabilistic neural language model. $C(w_i)$ is a word vector corresponding to word $i$ [21].

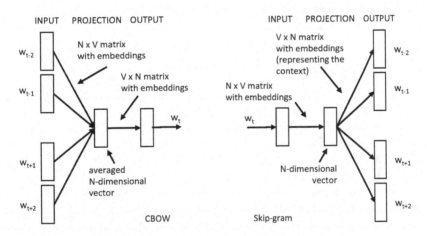

**Figure 13.4:** The architectures of neural networks for training word embeddings using the CBOW and Skip-gram approaches. $V$ is the size of the vocabulary, $N$ is the size of word vectors, $w_i$ are one-hot encoded vectors of size $V$ containing indices of the given words [195, 276].

calculated from a term-term matrix created from a corpus. The prediction is made by a function that takes word vectors as the input. The word vectors are calculated in the process of word co-occurrence matrix factorization using stochastic gradient descent which is a process that can be parallelized [214]. Compared to *word2vec*, the training is faster but it requires more memory.

■ **fastText.** Most of the techniques for learning word embeddings ignore sub-word information (prefixes, suffixes) which is very important for morphologically rich languages with many rare words. The *fastText* model, which is derived from *word2vec*, treats each word as a bag of character n-grams where the vectors are associated at the n-gram level. The vector for a word is calculated as the sum of n-gram vectors. This also enables creating vectors for words that are not in the training data [28].

When generating embeddings is a part of a neural network, the vectors can be initialized randomly as in the case of other model parameters. The values can be sampled uniformly from a fixed interval centered at zero (for example, the Xavier method samples the values from the interval $\left[ -\frac{\sqrt{6}}{\sqrt{n_i + n_o}}, +\frac{\sqrt{6}}{\sqrt{n_i + n_o}} \right]$, where $n_i$ and $n_o$ are the sizes of input and output), or from a zero-mean normal distribution with the standard deviation varying from 0.001 to 10 [148].

Instead of training, word vectors embeddings pre-trained on many different collection can be found. The collections are trained on sources like Wikipedia articles, tweets, Common Crawl, Google news and others.

## 13.5 Aggregation of Word Vectors

To represent larger pieces of text (sentences, paragraphs, documents), word vectors need to be aggregated somehow. Otherwise, the text can be viewed as a bag of embedded words with a variable length [56]. Many algorithms, however, require fixed length vectors. The fixed length vectors can be an input to, for example, classification or clustering algorithms instead of document vectors from the bag-of-words model.

Simple aggregation approaches use, for example, the sum, average, or a weighted average (using tf-idf wights) of the word vectors [66, 195].

During training word vectors, it is generally expected that the vectors should help in predicting other words in the given context. A similar idea is used by Le and Mikolov [162] in their *Paragraph Vector* algorithm. A larger piece of text, here a paragraph, should be useful in predicting other words too. The paragraph vector then behaves like another word in the prediction task. It represents the missing information from the current context and acts as a memory of the topic of the paragraph. Word vectors are shared across all paragraphs while the paragraph vectors are not.

A recursive neural network architecture to aggregate words vectors was proposed in [253]. The method requires parsing and thus works on a sentence level.

Another method to aggregate and compress the variable-size word embedding sets to binary hash codes through Fisher kernel and hashing methods was proposed in [290].

## 13.6   An Example

To train a *word2vec* model, the `wordVectors` library can be used. At this moment, it is not in the CRAN archive. It is, therefore, necessary to download it from github using `devtools`.

```
library(devtools)
install_github("bmschmidt/wordVectors")

library(wordVectors)
```

The author of `wordVectors` demonstrates the functionality on texts from cookbooks from Michigan State University. The archive with cookbooks is downloaded, stored as `cookbooks.zip` and unzipped to directory `cookbooks`. The directory will contain more than 70 text files with some cookbooks.

```
download.file("http://archive.lib.msu.edu/
              dinfo/feedingamerica/cookbook_text.zip",
              "cookbooks.zip")
unzip("cookbooks.zip", exdir="cookbooks")
```

To train a model, a text file containing the texts is needed. This file can be prepared in may different ways. The `wordVectors` library provides function `prep_word2vec()` that determines a file or directory with the original texts (parameter `origin`), the output file name (`destination`), information about n-gram handling (parameter `bundle_ngrams` – if the value is greater than 1, joins together common n-grams into a single word), and the possibility to lowercase the text (`lowercase`).

```
prep_word2vec(origin="cookbooks",
              destination="cookbooks.txt",
              lowercase=TRUE,
              bundle_ngrams=2)
```

Subsequently, a *word2vec* model can be trained using the `train_word2vec()` function on the created data file. Training parameters include the size of word vectors (parameter `vectors`), the size of the window around a word (`window`), the number of threads to run the training process on (`threads`), specification about whether to use the CBOW or skip-gram approach ((`cbow`), the minimal frequency of a word in the corpus to be included in the samples (`min_count`), the number of passes over the corpus in training (`iter`), or the number of negative samples in the skip-gram approach (`negative_samples`). The word vectors are written to an output file with in binary form (parameter `destination`). The function returns an object containing word vectors.

```
model <- train_word2vec(train_file="cookbooks.txt",
                        output_file="cookbook_vectors.bin",
                        vectors=200,
                        threads=4,
                        window=12,
                        iter=5,
                        negative_samples=0)
```

Individual word vectors can be accessed through a string containing the word of interest. In the following example, we are using the function `closest_to()` to find words most similar to *steak* and *cheese* (two ingredients that are mentioned in cookbooks).

```
> x <- model["steak", ]
> closest_to(model, x)
                word similarity to x
1               steak       1.0000000
2           beefsteak       0.7224785
3          rump_steak       0.6747437
4   porterhouse_steak       0.6517441
5              steaks       0.6482365
6             sirloin       0.5996025
7      planked_sirloin       0.5941793
8          roast_beef       0.5836023
9          tenderloin       0.5827629
10              broil       0.5745636
>
> x <- model["cheese", ]
> closest_to(model, x)
                word similarity to x
1              cheese       1.0000000
2        swiss_cheese       0.6298522
3                edam       0.6113659
4         emmenthaler       0.5890561
5             stilton       0.5821470
6            cheshire       0.5780776
```

| 7  | gruy_232       | 0.5679422 |
|----|----------------|-----------|
| 8  | kase           | 0.5519617 |
| 9  | cottage_cheese | 0.5487433 |
| 10 | neufchatel     | 0.5484801 |

We can also try some vector arithmetic – we want to know what is to *pork* as *steak* is to *beef*.

```
> x <- model["steak", ] - model["beef", ] + model["pork", ]
> closest_to(model, x)
# this is also possible:
#  closest_to(model, ~"steak"-"beef"+"pork")
```

|    | word       | similarity to x |
|----|------------|-----------------|
| 1  | steak      | 0.7268982       |
| 2  | pork       | 0.6755440       |
| 3  | steaks     | 0.5458622       |
| 4  | bacon      | 0.5458076       |
| 5  | chops      | 0.5364155       |
| 6  | pork_chops | 0.5351922       |
| 7  | broil      | 0.5095040       |
| 8  | broiled    | 0.4904118       |
| 9  | beefsteak  | 0.4878030       |
| 10 | ham        | 0.4817139       |

Another example will reveal what can be made from apples as butter is made from milk.

```
> x <- model["butter", ] - model["milk", ] + model["apple", ]
> closest_to(model, x)
```

|    | word               | similarity to x |
|----|--------------------|-----------------|
| 1  | apple              | 0.6531934       |
| 2  | apples             | 0.5038533       |
| 3  | tart               | 0.4672218       |
| 4  | butter             | 0.4625596       |
| 5  | lamb's_sweetbreads | 0.4581400       |
| 6  | pie                | 0.4531101       |
| 7  | apple_marmalade    | 0.4487496       |
| 8  | 872                | 0.4362927       |
| 9  | sauce              | 0.4331859       |
| 10 | compote            | 0.4307663       |

The following two examples try to show some syntactic regularities (plural/singular and base/continuous form of a verb).

```
> x <- model["apples", ] - model["apple", ] + model["pear", ]
> closest_to(model, x, n=5)
            word similarity to x
1           pear           0.8229101
2          pears           0.6867127
3         apples           0.6411531
4        peaches           0.6117798
5       apricots           0.5911249

> x <- model["baking", ] - model["bake", ] + model["cook", ]
> closest_to(model, x)
            word similarity to x
1           cook           0.5258886
2        cooking           0.4665998
3       roasting           0.4510705
4        stewing           0.4445075
5         baking           0.4195846
```

*GloVe* vectors can be learned with the `text2vec` package. In our example, the same text data (cookbooks) will be used.

```
library(text2vec)

# all files with cookbooks
file_list <- list.files(path='cookbooks', pattern="*.txt")
for (f in file_list) {
    # creating a full name including file and directory names
    fullname <- paste("cookbooks", "/", f, sep="")
    # reading all lines from a file
    t <- readLines(fullname, n=-1)
    # appending the cookbook to all texts
    text <- append(text, t)
}

# creating tokens from the text
tokens <- word_tokenizer(text)

# creating an iterator over the list of character vectors
```

```
# which are the documents being processed
it <- itoken(tokens, progressbar=TRUE)

# creating the vocabulary using the iterator
v <- create_vocabulary(it)

# filtering out too rare words
v <- prune_vocabulary(v, term_count_min=3)

# a function mapping words to indices, it is later used
# to create a term co-occurrence matrix
vectorizer <- vocab_vectorizer(v)

# creating a term-co-occurrence matrix from windows
# of size 5 words to the left and right from a word
tcm <- create_tcm(it, vectorizer,
                  skip_grams_window=10,
                  skip_grams_window_context="symmetric")

# creating GloVe word vectors:
#  - creating an object for a GloVe model
glove <- GlobalVectors$new(word_vectors_size=300,
                           vocabulary=v,
                           x_max=10)
#  - fitting the model to a term-co-occurrence matrix
#    (10 iterations of stochastic gradient descent, stopping
#     when two subsequent iterations bring an improvement less
#     than convergence_tol); the model learns vectors for main
#     as well as for context words
vectors_main <- glove$fit_transform(tcm,
                                     n_iter=10,
                                     convergence_tol=0.01)
vectors_context <- glove$components

#  - the sum of main and context vectors is calculated
word_vectors <- vectors_main + t(vectors_context)
```

Now, we can query the model to see some regularities in the learned word vectors. To calculate similarities between rows of two matrices, function sim2() from package text2vec is used. The first parameter is the matrix with word vectors, the second is a row related to one word. To maintain the matrix type for the row, parameter drop is set to FALSE.

```r
# calculating cosine similarities to word "beef"
> cos_sim <- sim2(x=word_vectors,
+                 y=word_vectors["beef", , drop = FALSE],
+                 method="cosine",
+                 norm="l2")
# looking at five most similar words
> head(sort(cos_sim[, 1], decreasing=TRUE), 5)

     beef      veal      meat    mutton      pork
1.0000000 0.7188874 0.6961990 0.6873831 0.6791247

# finding out, what is to pork as steak is to beef
> y <- word_vectors["steak", , drop = FALSE] -
+      word_vectors["beef", , drop = FALSE] +
+      word_vectors["pork", , drop = FALSE]

> cos_sim <- sim2(x=word_vectors,
+                 y=y,
+                 method="cosine",
+                 norm="l2")
> head(sort(cos_sim[, 1], decreasing=TRUE), 5)

    steak      pork     bacon     chops      lard
0.7276466 0.6033111 0.3629336 0.3597848 0.3501749
```

# Chapter 14

# Feature Selection

## 14.1  Introduction

Many algorithms, like the ones for classification or clustering, work with structured representations of texts. These representations, especially in the bag-of-words form, suffer from many dimensions and sparsity. This generally decreases the performance of learning and the quality of achieved results [145]. It is therefore desirable to lower the number of features because it is expected that some of the features are irrelevant (removing them does not affect learning) or redundant (correlated with another feature) [175].

There are two possible approaches [179, 286, 159, 175]:

- ■ Feature extraction – new features are derived from the data through some mapping of features to others. Methods based on the decomposition of a matrix representing the distribution of terms in documents, like the Latent Semantic Analysis, is one of the examples. A problem with these techniques might be bad interpretability of the newly created features.

- ■ Feature selection – a subset from the original features is selected according to corpus statistics or some other criteria. The features can be ranked and the top $k$ or a minimal set that still enables satisfactory learning is selected.

In both approaches, it is desirable to select or extract the features automatically.

The main benefits of feature selection include [69, 201, 92]:

- ■ Generalization of the model is better. This reduces overfitting and enables better learning accuracy.

■ Efficiency of the algorithms increases. The time needed for computations, space for storing data, network bandwidth to transmit data can be decreased for the purpose of data collection, training, and execution.

■ The crated model can be more easily understood, visualized (although it might not be possible to reduce the features to two or three dimensions so that the model can be examined by a human), and interpreted by a researcher.

Supervised feature selection methods rely on the knowledge of class labels of the data items. Then, the features that are somehow related to the labels are relevant. In an unsupervised problem, the class labels are missing, so feature selection is more complicated, but still desirable. The impact of feature selection needs to be evaluated differently. In a supervised selection, the features should be able to distinguish between the classes of data items, and common classification performance metrics, like accuracy, can be used. In unsupervised feature selection, the features should help reveal interesting groups in data. Clustering evaluation measure can be used here. The problem is that the number of clusters is usually not known in advance and also differs for different numbers of features (dimensions in the feature space) [78].

The following approaches can generally be distinguished in feature selection. They can be used in supervised feature selection, as well as in unsupervised feature selection [47, 159, 137, 78, 119]:

■ Wrapper approaches – a subset of features is evaluated by an algorithm (e.g., a classification algorithm) and by an evaluation metrics of that algorithm (e.g., classification accuracy). Because a new learner needs to be created for each subset, this approach might be very computationally demanding. On the other hand, this approach can lead to better performance as compared to a filter approach.

■ Filter approaches – a subset of features is selected independently on an algorithm to be later applied to the data. It generally considers some discriminative ability of features, e.g., the ability to distinguish classes.

■ Embedded approaches – a feature selection process is automatically embedded in an algorithm. An example is the selection of a feature for split in decision tree learning (e.g., the C4.5 algorithm [220]) or W-k-means, a modified k-means algorithm which automatically determines important features [121].

■ Hybrid approaches – take advantage of both filter and wrapper approaches. An initial set of features is selected with computationally efficiat filters and then refined by more accurate wrappers.

## 14.2 Feature Selection as State Space Search

In order to find the best features for a given problem, all possible feature subsets should be examined. This is usually not feasible, so some search strategies are applied. To introduce some system to the process, a procedure systematically adding or removing features is usually applied. In other words, the state described by the features selected or unused for the given problem changes in a systematic manner. There are a certain number of possible states and transitions between them. The goal is to find a suitable path from an initial state to a state that satisfies given conditions (e.g., the features enable achieving small classification error). Several approaches to the definition of the states and transitions between them are in existence [27]:

■ Forward selection – the feature set is empty at the beginning. The features are successively added until a suitable feature set is found. This can be seen as searching a state space where the states are described by selected and unused features; see Figure 14.1.

■ Backward elimination – the feature set contains all features at the beginning. The features are successively eliminated as long as the features are useful. The states in the state space defined by selected and unused attributes are searched in the order inverse to the search in forward selection.

In order to visit all states during searching for a suitable feature set, $2^m$ possible states need to be visited ($m$ is the number of features). In the text mining domain, where the data can be characterized by tens of thousands of features, this is sometimes not realistic. Instead of exhaustive state space search (which would guarantee finding an optimal feature set), some greedy approach that considers local changes between the states is applied. These heuristic approaches are, besides being less computationally demanding, often less prone to overfitting [175].

The search strategies for traversing the state space include [137, 258]:

■ Exponential methods – The number of feature subsets grows exponentially with the increasing feature space size. An example is the exhaustive search.

■ Sequential methods – With the growing feature space size, only one or few subsets need to be evaluated. Greedy forward selection or backward elimination belong to this category.

■ Randomized methods – Some of the steps in selecting features are randomized (for example, randomly selecting a subset of features). This usually increases the speed of computations and prevents trapping in a local optimum.

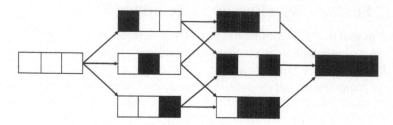

**Figure 14.1:** Forward feature selection strategy. We start with en empty feature set and search the state space defined by the set of selected/unused features by adding additional features. At the end, all features are selected. If the arrows are inverted and we start with the full feature set, we talk about backward elimination.

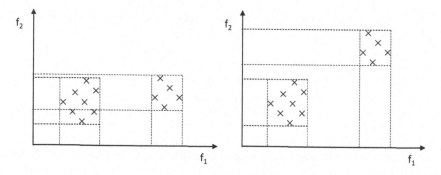

**Figure 14.2:** Left: feature $f_1$ is able to separate the objects to two clusters. Feature $f_2$ is useless in this task, it is irrelevant. Right: both features can separate the objects to two clusters, but only one of them is enough. They contain redundant information [78].

## 14.3   Feature Selection Methods

Wrapper methods often work better than filters because they are tailored to a specific problem. Their application is, however, often limited to smaller problems. On the other hand, filter methods are preferred for large scale problems, like text categorization, because of their lower computational costs [169].

Commonly used filter approaches to feature selection in text mining include chi squared, mutual information, information gain, bi-normal separation (BNS) expected cross entropy, Gini index, odds ratio, term strength, entropy-based ranking, and term contribution [200, 137, 92].

Supervised filter methods rank attributes according to their utility with respect to a given task (e.g., assigning a document to a class). This utility is assigned by a function that Li et al. [169] call a *score-computing function*. Such functions, for a supervised task, usually work with probabilities that are estimated based on statistical information about distribution of terms in documents with different categories. These probabilities include:

■ $P(t_i)$: the probability that a document contains term $t_i$

■ $P(\bar{t}_i)$: the probability that a document does not contain term $t_i$

■ $P(c_j)$: the probability that a document belongs to category $c_j$

■ $P(\bar{c}_j)$: the probability that a document does not belong to category $c_j$

■ $P(t_i, c_j)$: the joint probability that a document contains term $t_i$ and also belongs to category $c_j$

■ $P(c_j|t_i)$: the probability that a document belongs to category $c_j$ given that it contains term $t_i$

■ $P(c_j|\bar{t}_i)$: the probability that a document belongs to category $c_j$ given that it does not contain term $t_i$

■ $P(\bar{c}_j|t_i)$: the probability that a document does not belong to category $c_j$ given that it contains term $t_i$

■ $P(t_i|c_j)$: the probability that a document contains term $t_i$ given that it belongs to category $c_j$

■ $P(t_i|\bar{c}_j)$: the probability that a document contains term $t_i$ given that it does not belong to category $c_j$

■ $P(\bar{t}_i|c_j)$: the probability that a document does not contain term $t_i$ given that it belongs to category $c_j$

These probabilities can be only estimated with some statistical information from the training data:

■ $A_j$: the number of documents that contain term $t_i$ and also belong to category $c_j$

■ $B_j$: the number of documents that contain term $t_i$ but do not belong to category $c_j$

■ $N_j$ : the number of documents that belong to category $c_j$

■ $N$ : the total number of documents

■ $C_j$ : the number of documents that do not contain term $t_i$ but belong to category $c_j$; can be calculated as $N_j - A_j$

■ $D_j$ : the number of documents that neither contain term $t_i$ nor belong to category $c_j$; can be calculated as $N - N_j - B_j$

For unsupervised methods, no information is available about class. The suitability of a feature with respect to a given task can thus be derived only from the data itself. Unsupervised methods, like term contribution, are sometimes criticized because they rely on a particular similarity function (typically the cosine similarity); see below. When a different similarity function would be used, different terms might be selected as relevant [7]. When many computations need to be performed, e.g., when calculating the similarity of all document pairs, sampling can be used in real experiments [179].

## 14.3.1 Chi Squared ($\chi^2$)

The Chi-squared test is used in statistics to measure independence of two events measured by two categorical variables. To test this independence, a random sample is selected, the properties of these objects are studied, and a contingency table with frequencies is created. We might, for example, study whether gender and selected majors in a university are related. Randomly picked 100 men and 100 women are asked to tell their major and the frequencies of specific gender–major pairs are counted (e.g., 20 men study *Computer Science*). The gender and major variables are then tested for their independence [264].

The test measures how what is observed is different from what is expected. If, for example, you toss a fair coin ten times, you expect that it will land five times on tails and five times on heads (tossing a coin eleven times, one would expect 5.5 tails and 5.5 heads, which, of course, is not realistic) [251].

The null hypothesis, which is tested, is that there is no statistically significant difference between the expected and observed frequencies [132]. To get the expected frequencies, row and column marginals are multiplied and then divided by the sum of all cells in the contingency table. Then, the $\chi^2$ value is calculated as the sum of the squares of differences between the observed and expected values divided by the expected values [93]:

$$\chi^2 = \sum \frac{(O-E)^2}{E}, \tag{14.1}$$

where $O$ is the observed frequency and $E$ is the expected frequency. The calculated value of $\chi^2$ is compared to the critical $\chi^2$ value (usually the $95th$ percentile of the $\chi^2$ distribution with one degree of freedom) to see how significant the difference between the observed and expected frequencies is [48].

The Chi-squared score can be used to measure the dependence between a term and category. The occurrences of the term and the class, both having values *yes* or *no* are the two events studied. The value of $\chi^2$ can be calculated as follows [185]:

$$\chi^2(t,c) = \sum_{e_t \in \{yes,no\}} \sum_{e_c \in \{yes,no\}} \frac{(O_{e_t e_c} - E_{e_t e_c})^2}{E_{e_t e_c}}, \tag{14.2}$$

where $O$ is the observed frequency, $E$ is the expected frequency, and $e_x$ is the number of documents containing or not containing term $t$ and belonging or not belonging to class $c$ (for example, $e_t = yes$, $e_c = yes$ is the number of documents containing term $t$ and belonging to class $c$).

Considering the statistics defined above, the value of $\chi^2$ can be calculated as follows [286]:

$$\chi(t,c) = \frac{N(A_jD_j - C_jB_j)^2}{(A_j + C_j)(B_j + D_j)(A_j + B_j)(C_j + D_j)} \tag{14.3}$$

If the term and category are independent, the value of $\chi^2$ is equal to zero. Higher values mean that some dependency exists between the class and the term. In other words, the occurrence of a term makes a class more or less likely.

To calculate the overall usefulness of a term (not just for one specific class), a weighted average over all classes is calculated [82]:

$$\chi(t) = \sum_{j=1}^{k} P(c_j)\chi^2(t,c_j), \tag{14.4}$$

where $P(c_j)$ is the probability of category $c_j$.

In text categorization, no statements about statistical independence need to be made and absolute values of $\chi^2(t,c)$ are not that important. Instead, the relative importance of features is interesting and the values of $\chi^2(t,c)$ are used to rank the terms. Useful terms that are able to distinguish classes have high values of $\chi^2(t,c)$ [185]. The values of $\chi^2$ are normalized so they can be used to compare terms in one category. However, this normalization is broken when there are low frequency terms [286].

## 14.3.2 Mutual Information

In the information theory, *information* is contained in a sequence of messages that are transmitted from a source to a receiver; it is expected to be useful in reducing uncertainty about an event. The content of the messages can be studied and quantified in terms of the element of uncertainty or surprise they bring. If we have a source sending only one possible message, we are pretty sure about its content. In other words, we are not surprised at all after seeing the message. If there are two equally possible outcomes, the chance of receiving one of them is 50%. One of 10 possible messages, equally likely to occur, the probability of receiving one of them will be 10%. We can see that with the growing number of possible outcomes, our uncertainty about the incoming message generally increases. The surprise is also related to the probability with which the different messages occur. Having two possible outcomes X and Y where X will be more likely to occur than Y, we will be more certain about the incoming message because we will see X more often.

To be able to measure uncertainty (surprise) of the outcome of a message, Shannon [249] defined information of a message, $x$, that comes from a set of all possible messages $X$ as:

$$I(x) = \log \frac{1}{p(x)} = -\log p(x), \qquad (14.5)$$

where $p(x)$ is the probability of message $x$. The base of the logarithm is often 2 because many things are boolean in nature. Messages are also often encoded using zeros and ones in computer memories. The unit of information is then known as a *bit* or *shannon*. When there are just two possible messages, like when tossing a fair coin, the information of a head will be $-\log_2 0.5 = 1$ bit. Having hexadecimal digits, the information associated with each of them equals to $-log_2 1/16 = 4$ bits.

The amount of information is related to the size of a message, e.g., in a communication channel. The value of $\frac{1}{p(x)}$ in the formula is actually the number of choices when the probabilities of each message are equal. To encode a coin toss, outcome, one binary digit is needed. To encode a hexadecimal digit, four binary digits are necessary. But what if we have a fair die? Then, each number on it will carry information $-log_2 1/6 = 2.58$ bits. It means that 2.58 binary digits are needed to encode a message. How we can use only a part of a binary digit? A better option is to think about asking questions allowing either yes or no answers that will subsequently divide all possible answers into two equally sized halves. Let us look at the example of guessing a number between 1 and 8 where, for instance, 7 is the right answer. The following questions might be asked:

■ Is the number less or equal to 4? No – the number must be 5, 6, 7, or 8.

■ Is the number greater or equal to 7? Yes – the number must be 7 or 8.

■ Is the number 8? No – the number must be 7.

We asked three questions to receive the correct answer, which equals to $-\log 1/8$. The more the questions which need to be asked, the more the information a message contains. We can look at the measure (of uncertainty) as the amount of missing information that is required to specify an outcome [164]. If the amount of information is not an integer number, we can treat is as the average number of questions to be asked to get a correct answer.

When some of the answers are more likely, it is better to ask questions so that their number is minimized (i.e., ask for the most likely answer). When relating this to encoding messages, it is better to encode more likely answers using a lower number of symbols (bits) because they will occupy less space. A well known example is the Morse code where more frequent letters, like E, T, A, or I have less symbols than, for example, Z, H, or Q. The Morse code is, however, not a good application for computers because it requires three symbols (dot, dash, space).

To encode letters of the English alphabet (26 symbols in total), $-\log_2 1/26 = 4.7$ bits, which means 5 bits of a fixed code would be needed. When considering the different frequency of the letters and, for example, a Huffman code, only 4.13 bits are needed on average (the most frequent letter, E, will be encoded as 000, A will become 0011, and the least frequent letter, Z, will be 1111111) [266, 233].

A measure of the average amount of information of a message with $n$ possibilities, when the probabilities $p_1, p_2, \ldots, p_n$ are taken into consideration, is known as *entropy*. Shannon [249] found a function $H(p_1, p_2, \ldots, p_n)$ that was the only one satisfying several important properties: is continuous in its parameters, is a monotonic increasing function of $n$ when $p_1 = p_2 = \ldots p_n = \frac{1}{n}$, and is the weighted sum of the individual values of $H$ when a choice is broken down into two successive choices.

The formula for calculating entropy, which is actually the weighted average of entropies of individual outcomes (weighted by their probability), is:

$$H = \sum_{i=1}^{n} p_i \log p_i \tag{14.6}$$

The logarithmic measure expressing the information of one message selected from several possible messages where all choices are equally likely, is the most natural choice because many important parameters in engineering vary linearly with the logarithm of the number of possibilities. When there are $N$ events with two possible outcomes, adding another such event will double the number of possible outcome combinations. This is expressed by the increase by 1 of the base 2 logarithm of the number of possible outcomes.

When there is only one possible outcome of an event, the value of entropy for this information is $H = -1 \log 1 = 0$. If the source sends two possible messages with the same probability (like, e.g., in the case of tossing a fair coin), the surprise can be quantified by entropy $h = -(0.5 \log_2 0.5 + 0.5 \log_2 0.5) = 1$ bit. Having two possible outcomes where one has a higher probability, say 70% (for example, an unfair coin flipping), the entropy will be $H = -(0.7 \log_2 0.7 + 0.3 \log_2 0.3) = 0.88$ bits. When there are more than two outcomes, for example, when rolling a fair die, the entropy will have the value $H = -6(1/6 \log_2 1/6) = 2.58$ bits. We can see that entropy (our surprise) is higher when there are more possible outcomes and when the outcomes have equal probability.

When there are two random variables $x$ and $y$, the *Pointwise Mutual Information* (PMI) can be calculated for two outcomes $x \in X$ and $y \in Y$ as follows [85]:

$$PMI(x, y) = \log \frac{p(x|y)}{p(x)} = \log \frac{p(x, y)}{p(x)p(y)} = \log \frac{p(y|x)}{p(y)} \tag{14.7}$$

Taking logarithm out of consideration, the formulae calculate the ratio of the probability when $x$ and $y$ co-occur and when they were independent. The logarithm is used to measure this value in bits (when the base is 2). It can be interpreted as the amount of information provided by the occurrence of $x$ about the occurrence of $y$ [85]. The value of PMI equals zero when the events $x$ and $y$ are independent, i.e., $p(x,y) = p(x)p(y)$. When there is an association between $x$ and $y$, the probability $p(x,y)$ will be greater than $p(x)p(y)$ and thus $PMI > 0$. If one event occurs while the other tends not to, which is known as complementary distribution, the value of PMI is negative [25, 52].

When considering all possible outcomes of two random variables $X$ and $Y$, *mutual information* is the average pointwise mutual information for all possible pairs $x \in X$ and $y \in Y$. It is, again, a weighted average where the weights are the probabilities that $x$ and $y$ co-occur:

$$I(X,Y) = \sum_{x \in X} \sum_{y \in Y} p(x,y) \log \frac{p(x,y)}{p(x)p(y)} \tag{14.8}$$

Mutual information measures how much information about one variable is contained in another variable.

When measuring joint entropy of two variables $X$ and $Y$ (uncertainty associated to co-occurrence of values of two variables), we measure average information for all possible outcomes $x \in X$ and $y \in Y$:

$$H(X,Y) = -\sum_{x \in X} \sum_{y \in Y} p(x,y) \log p(x,y) \tag{14.9}$$

The joint entropy $H(X,Y)$ is always non-negative, greater than or equal to $H(X)$ or $H(Y)$, and less than or equal to the sum of individual entropies:

$$H(X,Y) \geq 0$$

$$H(X,Y) \geq max[H(X),H(Y)] \tag{14.10}$$

$$H(X,Y) \leq H(X) + H(Y)$$

If the variables $X$ and $Y$ are independent, the information contained in their combination equals the sum of individual entropies. If the variables are somehow dependent, they provide less information in their combination. This smaller amount information is quantified by their mutual information:

$$I(X,Y) = H(X) + H(Y) - H(X,Y) \tag{14.11}$$

The joint entropy can also be related to the conditional entropy. The conditional entropy measures how much information is contained in one variable when the other is known. When we know the value of variable $X$, the conditional entropy $H(Y|X = x)$ can be calculated as follows:

$$H(Y|X = x) = -\sum_{y \in Y} p(y|x) \log p(y|x) \tag{14.12}$$

The conditional entropy $H(Y|X)$ is then an average of $H(Y|X = x)$ for all possible values of $x$:

$$
\begin{aligned}
H(Y|X) &= \sum_{x \in X} p(x) H(Y|X = x) \\
&= -\sum_{x \in X} p(x) \sum_{y \in Y} p(y|x) \log p(y|x) \\
&= -\sum_{x \in X} \sum_{y \in Y} p(x,y) \log p(y|x)
\end{aligned}
\tag{14.13}
$$

Mutual information can also be interpreted as the information contained in one variable minus the information contained in the variable when the other variable is known [104]:

$$
I(X,Y) = H(Y) - H(Y|X) = H(X) - H(X|Y)
\tag{14.14}
$$

Mutual information can be used to measure how a class $c \in C$ and term $t \in V$ occur together, in other words, how much information about the classes the terms carry. An ideal situation would be achieved when a subset $s \subset V$ has the same mutual information with $C$ as $V$ has with $C$:

$$
I(S,C) = I(V,C)
\tag{14.15}
$$

Because this problem is intractable, features with high mutual information with $C$ are included in $S$ so $I(S,C)$ is close to $I(V,C)$ [176].

To evaluate how a term $t$ and class $c$ are correlated, their mutual information can be estimated as follows:

$$
I(t,c) = \log \frac{AN}{(A+C)(A+B)}
\tag{14.16}
$$

In order to measure the contribution of a term in assigning a class globally, there are a few ways in which the term-category scores can be combined into one score [286]:

$$
I_{avg}(t) = \sum_{i=1}^{m} p(c_i) I(t,c_i) I_{max}(t) = \max i = 1^m [I(t,c_i)]
\tag{14.17}
$$

A weakness of the mutual information measure is that it assigns higher scores to rare terms that have the same conditional probability $p(t|c)$ as common terms [286].

### 14.3.3 Information Gain

Values of some features are related to specific classes and it is expected that instances from different classes will have different values of these features. For example, body height is expected to be higher for adults and lower for children. If such a relationship is ideal, after partitioning all instances to be classified according to the value of a specific feature, we will have groups where there are instances of only one class in each group (e.g., in a group where body height is high there will only be adults, in the other group where body height is low only children will be present). We can say, that after partitioning one heterogeneous group of instances (with several mixed classes), we get several sets of homogeneous data (with only one class in a group).

The concept of entropy can be used to quantify how well a feature contributes to such partitioning. Entropy is used to measure the surprise of seeing an instance of a specific class in a set. The original heterogeneous set contains instances of several classes and the data is thus characterized by a certain level of entropy which is rather high. After having several more homogeneous sets where instances of one class prevail, the level of entropy will be lower. The feature thus contributes to the elimination of entropy that measures a disorder in the data.

Reduction of entropy (uncertainty) is related to information which needs to be sent to describe the data. When the data is heterogeneous, more information is needed than when there are several homogeneous sets. We thus gain some information that does not need to be sent. The measure quantifying such a contribution of an attribute is known as *information gain*.

The value of information gain for term $t_i$ can be calculated as the difference between the entropy of the set containing all data and the average entropy of sets that are the result of splitting the data according to the values of attribute $t_i$ [199]:

$$IG(t_i) = Entropy(S) - \sum_{v \in values(t_i)} \frac{|S_v|}{|S|} Entropy(S_v), \qquad (14.18)$$

where $S$ is the set containing all data, $values(t_i)$ is a set of all possible values of feature $t_i$, and $S_v$ is a subset of $S$ where the value of $t_i$ is $v$. The entropy of data after partitioning is calculated as the weighted average of individual entropies where the weights are the relative sizes of the sets.

When we consider only presence or absence of a feature (i.e., presence or absence of a word in a text) information gain can be calculated as

$$
\begin{aligned}
IG(t_i) = &-\sum_{j=1}^{m} p(c_j) \log p(c_j) \\
&+ p(t_i) \sum_{j=1}^{m} p(c_j|t_i) \log p(c_j|t_i) \\
&+ p(\overline{t_i}) \sum_{j=1}^{m} p(c_j|\overline{t_i}) \log p(c_j|\overline{t_i}),
\end{aligned}
\qquad (14.19)
$$

where $m$ is the number of classes, and $p(t_i)$ and $p(\bar{t}_i)$ are the probabilities that term $t_i$ appears or does not appear in a document.

The value of information gain can be estimated as follows [169]:

$$
\begin{aligned}
IG(t_i) = {} & - \sum_{j=1}^{m} \frac{N_j}{N_{all}} \log \frac{N_j}{N_{all}} \\
& + \sum_{j=1}^{m} \frac{A_j}{N_{all}} \sum_{j=1}^{m} \frac{A_j}{A_j + B_j} \log \frac{A_j}{A_j + B_j} \\
& + \sum_{j=1}^{m} \frac{C_j}{N_{all}} \sum_{j=1}^{m} \frac{C_j}{C_j + D_j} \log \frac{C_j}{C_j + D_j}
\end{aligned}
\tag{14.20}
$$

Information gain is an important metric for determining the suitability of algorithms in the induction of decision trees with the *c4.5* algorithm; see [7]. The method is actually embedded in the algorithm.

## 14.4   Term Elimination Based on Frequency

Frequency based approaches to feature selection do not usually perform as well as other, more sophisticated methods, but often bring acceptable results [175]. They are based on a simple idea that the frequency with which a term appears is related to its importance.

Terms appearing too frequently usually do not have a discriminative ability. Words like *the*, *in*, or *have* are contained in almost every document and are not specific for one class of documents. They are also not able to clearly characterize a group of documents in a clustering problem for the same reason. The common terms are known as stop words. Many lists of commonly used stop words are available in many languages. They usually contain about 300–400 terms [7].

Rare terms, on the other hand, can distinguish classes in classification. If a term appears in only one document from a specific class, one can think about using it when deciding on a class label. However, the goal of a learning algorithm is to find a generalization of class label assignment. A decision based on only one evidence would not be very general and will most likely lead to overfitting. Rare terms do not contribute significantly to calculating similarity in either. These rare terms are often noise in the data and represent mistyping or other language errors.

The rare and common terms can be easily eliminated without significant impact on the performance of a learner in a classification or clustering problem [270, 64]. In an information retrieval task, rare terms are considered to be informative. Thus, strong reduction of these terms is not recommended [286].

The frequencies used to determine a term's importance can be document frequencies or the total frequencies. Document frequency is more suitable for the

Bernoulli model (when the presence or absence of a term in a document is important), the total frequency is better for the multinomial model (when we have information about the frequency of a term in a document) [185].

Another possibility is to work only with terms that are most common in a class. Of course, many stop words bringing no specific information of the class will be included but when a sufficient number of features is used, the results are acceptable [185].

## 14.5   Term Strength

The Term Strength measure was originally proposed by Wilbur and Sirotkin [278] to automatically discover stop words in information retrieval. The term strength, or importance, is based on the occurrence of a term in related documents. If the labels of documents are known, related documents belong to the same category. If the labels are unknown, the relatedness can be defined by a human. This is, however, often impractical so the relatedness is calculated using a document similarity, typically the Cosine similarity (it is expected that related documents share some words and the similarity is high). It is necessary to define a minimum threshold level for the similarity in order to consider two documents be related.

The term strength measures the probability with which a term occurs in the second document in a highly related pair of documents, knowing that it also occurs in the first one:

$$s(w) = p(w \in d_j | w \in d_i), d_i \neq d_j \tag{14.21}$$

To estimate the term strength, we can randomly pick pairs of documents from the entire collection and calculate:

$$s(w) = \frac{number\ of\ pairs\ where\ w\ occurs\ in\ d_i\ and\ d_j}{number\ of\ pairs\ where\ w\ occurs\ in\ d_i} \tag{14.22}$$

A term is expected to be a stop word (thus, an irrelevant word) if its term strength significantly differs from the term strength of a random term (a term randomly distributed in training documents with the same frequency). The significant difference means that the actual term strength is less than or equal to the term strength of a random word (expected term strength) plus twice the standard deviation of this term strength [7]. To see the details regarding the calculation of the expected term strength of a random term and its random deviation, see [278].

## 14.6   Term Contribution

Term Contribution was proposed by Liu et al. [179] for selecting relevant terms for clustering. In clustering, similarity of documents is of great importance, so Term Contribution measures how a term contributes to document similarity. When calculating the similarity, the cosine similarity is considered as the similarity measure. The similarity of two documents, $d_i$ and $d_j$, is calculated as the dot product of the length normalized feature vectors representing the documents:

$$sim(d_i, d_j) = \sum_t w(t, d_i) \times w(t, d_j), \qquad (14.23)$$

where $w$ represents a normalized tf-idf weight of term $t$ in document $d_i$ or $d_j$. In a $d_i$ or $d_j$ document pair, term $t_t$ contributes to the dot product (i.e., the similarity) with $w(t_t, d_i) \times w(t_t, d_j)$. Term contribution of the term is calculated as the sum of these contributions over all pairs in the document collection $D$:

$$TC(t_t) = \sum_{d_i, d_j \in D, d_i \neq d_j} w(t, d_i) \times w(t, d_j) \qquad (14.24)$$

Terms with high scores of Term Contribution are considered to be relevant.

## 14.7   Entropy-Based Ranking

This measure is based on distinguishing between data with and without clusters. When clusters exist, objects in a cluster are expected to be close to each other in a feature space. Some of the features contribute to partitioning objects into clusters more than others and are thus important. Their usefulness is based on the idea that, after removing them, clustering will not be that obvious.

The distribution of objects in a feature space, in other words a disorder in the space, can be measured by entropy. If the probability of each point is equal, in other words, the objects are uniformly distributed, entropy is maximal.

To estimate the entropy (normalized to the interval $[0, 1]$) of a collection of $n$ documents, similarities $S_{ij}$ between all pairs of objects $X_i$ and $X_j$ are used in calculations [63]:

$$E = -\sum_{i=1}^{n} \sum_{j=1}^{n} [S_{ij} \log S_{ij} + (1 - D_{ij}) \log(1 - D_{ij}] \qquad (14.25)$$

The quality of a term is calculated as an entropy reduction when a term is removed from the collection [7].

## 14.8 Term Variance

The Term Variance measure proposed by Liu et al. [178] is based on the simple idea that important terms have high document frequencies and are not distributed uniformly. That means that they appear with higher frequencies in some documents and with lower frequencies in other documents. Terms appearing in just a few documents or having low variance will have a low Term Variance value. Term Variance of term $w_i$ in a collection of $n$ documents can be calculated as follows:

$$TV(w_i) = \sum_{j=1}^{n} \left( f_{ij} - \overline{f}_i \right)^2 , \qquad (14.26)$$

where $f_{ij}$ is the frequency of term $w_i$ in $j^{th}$ document and $\overline{f}_i$ is the average frequency of $w_i$ in documents.

## 14.9 An Example

In the example, consider that we have 30 documents in three categories (*sport*, *politics*, and *technology*). We look only at a few terms (*today, and, football, hockey, poll, minister, meeting, computer,* and *learning*) and their presence in the documents in the three classes. A document-term matrix, together with document class labels representing the newspaper articles, can be found in Figure 14.9.

A summary containing the distribution of terms in the three categories is in Figure 14.4. We can see that the words are distributed differently in individual classes. Some of them, like the word *today*, appear with almost the same frequency in all classes; others, like *hockey* or *learning*, seem to be more specific for only some categories.

To quantify the importance of terms in the task of assigning a class label, the values of some feature selection metrics can be calculated.

Some already implemented feature selection algorithms are contained, for example, in package FSelector. The package provides algorithms for filtering attributes (e.g., functions chi.squared(), information.gain()), for wrapping classifiers and searching the attribute subset space using backward and forward search, and for choosing a subset of attributes.

In the example, a document-term matrix is created and passed to one method (information gain) of feature selection from package FSelector. Several functions of the package are used to show the most important features and to remove the unimportant features from the data.

| today | and | football | hockey | poll | minister | meeting | computer | learning | CLASS |
|---|---|---|---|---|---|---|---|---|---|
| 1 | 1 | 0 | 1 | 0 | 0 | 1 | 0 | 0 | sport |
| 0 | 0 | 1 | 0 | 0 | 0 | 0 | 0 | 0 | sport |
| 1 | 0 | 1 | 1 | 0 | 0 | 0 | 0 | 0 | sport |
| 0 | 1 | 0 | 0 | 0 | 1 | 0 | 1 | 0 | sport |
| 0 | 0 | 0 | 1 | 0 | 0 | 0 | 0 | 1 | sport |
| 1 | 0 | 1 | 0 | 1 | 0 | 1 | 0 | 0 | sport |
| 1 | 1 | 1 | 0 | 0 | 0 | 1 | 0 | 0 | sport |
| 0 | 0 | 1 | 1 | 0 | 0 | 0 | 0 | 1 | sport |
| 1 | 0 | 0 | 1 | 0 | 0 | 0 | 0 | 0 | sport |
| 0 | 1 | 1 | 0 | 0 | 1 | 0 | 1 | 0 | sport |
| 0 | 0 | 1 | 0 | 0 | 0 | 1 | 0 | 0 | politics |
| 1 | 1 | 0 | 0 | 1 | 1 | 0 | 0 | 0 | politics |
| 0 | 0 | 0 | 0 | 0 | 1 | 0 | 0 | 1 | politics |
| 0 | 1 | 0 | 0 | 0 | 0 | 1 | 0 | 0 | politics |
| 1 | 1 | 0 | 1 | 0 | 0 | 0 | 0 | 0 | politics |
| 1 | 1 | 0 | 0 | 1 | 0 | 0 | 0 | 0 | politics |
| 1 | 0 | 1 | 0 | 0 | 1 | 1 | 0 | 1 | politics |
| 0 | 1 | 0 | 0 | 1 | 0 | 0 | 1 | 0 | politics |
| 1 | 0 | 0 | 0 | 0 | 1 | 0 | 0 | 0 | politics |
| 1 | 1 | 0 | 0 | 1 | 0 | 1 | 0 | 0 | politics |
| 1 | 0 | 1 | 0 | 0 | 0 | 0 | 0 | 1 | technology |
| 0 | 1 | 1 | 0 | 0 | 0 | 0 | 0 | 1 | technology |
| 1 | 1 | 0 | 1 | 0 | 0 | 1 | 0 | 0 | technology |
| 0 | 1 | 0 | 0 | 0 | 1 | 0 | 1 | 0 | technology |
| 1 | 0 | 1 | 0 | 0 | 0 | 0 | 0 | 1 | technology |
| 1 | 0 | 0 | 0 | 0 | 0 | 0 | 0 | 0 | technology |
| 1 | 1 | 0 | 0 | 0 | 1 | 0 | 1 | 1 | technology |
| 1 | 1 | 0 | 0 | 1 | 0 | 1 | 0 | 0 | technology |
| 0 | 1 | 0 | 1 | 0 | 0 | 0 | 0 | 1 | technology |
| 0 | 0 | 0 | 0 | 0 | 0 | 0 | 1 | 0 | technology |

**Figure 14.3:** A term document matrix representing the distribution of nine terms in thirty documents from three classes. The matrix also contains class labels.

The values of another method (Chi squared) are calculated using coefficients A, B, C, and D, introduced earlier in this chapter. The most important features characterizing the classes are also presented. The features that have high values of Chi squared are shown here. A feature does not have to be present in documents from a class in order to have a high value (if it is not there, it also demonstrates some correlation between the feature and class). Thus, only features that appear in the classes with higher than average frequency are displayed.

|  | sport | politics | technology |
|---|---|---|---|
| today | 5 | 6 | 6 |
| and | 4 | 6 | 6 |
| football | 6 | 2 | 3 |
| hockey | 5 | 1 | 2 |
| poll | 1 | 4 | 1 |
| minister | 2 | 4 | 2 |
| meeting | 3 | 4 | 2 |
| computer | 2 | 1 | 3 |
| learning | 2 | 2 | 5 |

**Figure 14.4:** Frequencies of terms across three categories of newspaper articles.

```
# a vector to be converted to a matrix representing
# the distribution of terms in documents
d <- c(
       1,1,0,1,0,0,1,0,0,
       0,0,1,0,0,0,0,0,0,
       1,0,1,1,0,0,0,0,0,
       0,1,0,0,0,1,0,1,0,
       0,0,0,1,0,0,0,0,1,
       1,0,1,0,1,0,1,0,0,
       1,1,1,0,0,0,1,0,0,
       0,0,1,1,0,0,0,0,1,
       1,0,0,1,0,0,0,0,0,
       0,1,1,0,0,1,0,1,0,
       0,0,1,0,0,0,1,0,0,
       1,1,0,0,1,1,0,0,0,
       0,0,0,0,0,1,0,0,1,
       0,1,0,0,0,0,1,0,0,
       1,1,0,1,0,0,0,0,0,
       1,1,0,0,1,0,0,0,0,
       1,0,1,0,0,1,1,0,1,
       0,1,0,0,1,0,0,1,0,
       1,0,0,0,0,1,0,0,0,
       1,1,0,0,1,0,1,0,0,
       1,0,1,0,0,0,0,0,1,
       0,1,1,0,0,0,0,0,1,
       1,1,0,1,0,0,1,0,0,
       0,1,0,0,0,1,0,1,0,
       1,0,1,0,0,0,0,0,1,
```

```
        1,0,0,0,0,0,0,0,0,
        1,1,0,0,0,1,0,1,1,
        1,1,0,0,1,0,1,0,0,
        0,1,0,1,0,0,0,0,1,
        0,0,0,0,0,0,0,1,0
     )

# creating a document-term matrix
dtm <- matrix(d, nrow=30, ncol=9, byrow=TRUE)

# adding column and row names
colnames(dtm) <- c("today","and","football",
                   "hockey","poll","minister",
                   "meeting","computer","learning")
class_labels <- rep(c("sport","politics","technology"),
                    each=10)

# visualization of the distribution of words in classes
tab <- t(rowsum(dtm, class_labels))
barplot(tab,
        horiz=TRUE,
        ylab="Category",
        xlab="Numbers of articles containing the word",
        legend=TRUE,
        args.legend=list("ncol"=5, "x"=15, "xjust"=0.5,
                         "y"=4, "yjust"=0.2)
)

# using a library to calculate feature selection metrics
library(FSelector)

# creating a data frame for the methods of FSelector
df <- data.frame(dtm, class_labels)

# converting numeric values to discrete values
df[sapply(df, is.numeric)] <-
        lapply(df[sapply(df,is.numeric)], as.factor)

# calculating importance of attributes using
# the information gain method
attrs <- information.gain(class_labels~., data)
print("Values of Information gain:")
print(attrs)
```

```
# selecting 3 best attributes
print("Three best attributes:")
print(cutoff.k(attrs, 3))

# selecting 50% best attributes
print("50% best attributes:")
print(cutoff.k.percent(attrs, 0.5))

# preserving only 50% most important attributes
data_reduced <- data[, cutoff.k.percent(attrs, 0.5)]

# calculating the values of feature selection metrics
# using coefficients A, B, C, and D
D<-C<-B<-A<-matrix(nrow=length(unique(class_labels)),
                   ncol=dim(dtm)[2])
rownames(D)<-rownames(C)<-rownames(B)<-rownames(A)<-
        unique(class_labels)
colnames(D)<-colnames(C)<-colnames(B)<-colnames(A)<-
        colnames(dtm)

for (c in unique(class_labels)) {
   for (w in c(1:dim(dtm)[2])) {
      A[c,w] <- sum(dtm[class_labels == c, w] != 0)
      B[c,w] <- sum(dtm[class_labels != c, w] != 0)
      C[c,w] <- sum(dtm[class_labels == c, w] == 0)
      D[c,w] <- sum(dtm[class_labels != c, w] == 0)
   }
}

# calculating the values of the Chi square metric
chi <- (dim(dtm)[1]*((A*D)-(C*B))^2)/((A+C)*(B+D)*(A+B)*(C+D))

# calculating the average frequency of terms in the collection
avg_freq <- colMeans(tab)

# calculating the average frequency of terms in each class
numbers_of_documents_in_clusters <- table(class_labels)
tab <- tab/as.vector(numbers_of_documents_in_clusters)

# printing attributes for each class sorted according to their
# importance represented by the value of Chi squared
print("The values of Chi square for attributes and classes:")
for (c in unique(class_labels)) {
```

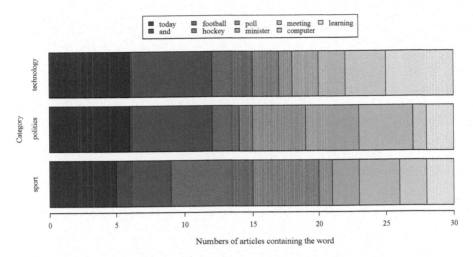

**Figure 14.5:** Visualization of term distribution across classes.

```
print(paste("Class: ", c))
# we want only attributes with the average frequency for
# a class higher than the average for all classes)
print(sort((chi[c,tab[,c ] > avg_freq]), decreasing=TRUE))
}
```

The output of the statements follows:

```
[1] "Values of Information gain:"
        attr_importance
today        0.00450826
and          0.01791164
football     0.06239830
hockey       0.07370431
poll         0.05934322
minister     0.02197633
meeting      0.01610484
computer     0.02161919
learning     0.04621363

[1] "Three best attributes:"
[1] "hockey"   "football" "poll"
[1] "50% best attributes:"
[1] "hockey"   "football" "poll"     "learning"
```

```
[1] "The values of Chi square for attributes and classes:"
[1] "Class:  sport"
 hockey football
4.176136 3.516746
[1] "Class:  politics"
     poll  minister   meeting        and      today
3.7500000 1.3636364 0.7142857 0.2678571 0.0678733
[1] "Class:  technology"
 learning  computer        and      today
2.8571429 0.9375000 0.2678571 0.0678733
```

# References

[1] S. Abney. *Semisupervised Learning for Computational Linguistics*. Chapman and Hall/CRC, 2007.

[2] S. P. Abney. Parsing by chunks. In R. C. Berwick, S. P. Abney, and C. Tenny, editors, *Principle-Based Parsing: Computation and Psycholinguistics*, pages 257–278. Kluwer, Dordrecht, 1991.

[3] M. Ackerman and S. Ben-David. A characterization of linkage-based hierarchical clustering. *Journal of Machine Learning Research*, 17:1–17, 2016.

[4] J. Adler. *R in a Nutshell: A Desktop Quick Reference*. O'Reilly, Sebastopol, CA, 2010.

[5] C. C. Aggarwal. An introduction to cluster analysis. In C. C. Aggarwal and C. K. Reddy, editors, *Data Clustering: Algorithms and Applications*, pages 1–28. CRC Press, Boca Raton, FL, 2014.

[6] C. C. Aggarwal. *Machine Learning for Text*. Springer International Publishing, Cham, 2018.

[7] C. C. Aggarwal and C. X. Zhai. A survery of text clustering algorithms. In C. C. Aggarwal and C. X. Zhai, editors, *Mining Text Data*, pages 77–128. Springer, New York, 2012.

[8] A. Albalate and W. Minker. *Semi-Supervised and Unsupervised Machine Learning: Novel strategies*. ISTE, Wiley, London, Hoboken, NJ, 2011.

[9] R. M. Aliguliycv. Clustering of document collection – a weighting approach. *Expert Systems with Applications*, 36(4):7904–7916, 2009.

[10] J. J. Allaire, F. Chollet, RStudio, and Google. https://keras.rstudio.com/. Accessed May 1, 2019.

[11] K. Allan, J. Bradshaw, G. Finch, K. Burridge, and G. Heydon. *The English Language and Linguistics Companion*. Palgrave Macmillan, New York, 2010.

[12] A. Amelio and C. Pizzuti. Correction for closeness: Adjusting normalized mutual information measure for clustering comparison. *Computational Intelligence*, 33(3):579–601, 2017.

[13] E. Amigó, J. Gonzalo, J. Artiles, and F. Verdejo. A comparison of extrinsic clustering evaluation metrics based on formal constraints. *Information Retrieval*, 12(4):461–486, 2009.

[14] G. Arakelyan, K. Hambardzumyan, and H. Khachatrian. Towards joint-tUD: Part-of-speech tagging and lemmatization using recurrent neural networks. In *Proceedings of the CoNLL 2018 Shared Task: Multilingual Parsing from Raw Text to Universal Dependencies*, pages 180–186. Association for Computational Linguistics, 2018.

[15] M. Bacchin, N. Ferro, and M. Melucci. A probabilistic model for stemmer generation. *Information Processing & Management*, 41(1):121–137, 2005.

[16] S. Bandyopadhyay and S. Saha. *Unsupervised Classification: Similarity Measures, Classical and Metaheuristic Approaches, and Applications*. Springer, Berlin Heidelberg, 2013.

[17] A. Barrón-Cedeño, C. Basile, M. D. Esposti, and P. Rosso. Word length n-grams for text re-use detection. In *Proceedings of the CICLING 2010 Conference*, volume 6008 of *LNCS*, pages 687–699, Berlin, Heidelberg, 2010. Springer.

[18] S. Basu and I. Davidson. Constrained partitional clustering of text data: an overview. In A. Srivastava and M. Sahami, editors, *Text Mining: Classification, Clustering, and Applications*, pages 155–184. CRC Press, Boca Raton, 2001.

[19] T. Bayes. An essay towards solving a problem in the doctrine of chances. *Philosophical Transactions of the Royal Society of London*, 53:370–418, 1763. http://www.stat.ucla.edu/history/essay.pdf.

[20] L. Belbin, D. P. Faith, and G. W. Milligan. A comparison of two approaches to betaflexible clustering. *Multivariate Behavioral Research*, 27:417–433, 1992.

[21] Y. Bengio, R. Ducharme, P. Vincent, and C. Janvin. A neural probabilistic language model. *Journal of Machine Learning Research*, 3:1137–1155, 2003.

[22] M. W. Berry, editor. *Survey of Text Mining Clustering, Classification, and Retrieval*. Springer, 2004.

[23] P. Bessiere, E. Mazer, J. M. Ahuactzin, and K. Mekhnacha. *Bayesian Programming*. Machine Learning & Pattern Recognition. Chapman & Hall/CRC, 1st edition, 2013.

[24] M. J. Best. *Quadratic Programming with Computer Programs*. Advances in Applied Mathematics. Chapman and Hall/CRC, 2017.

[25] D. Biber, S. Conrad, and R Reppen. *Corpus Linguistics: Investigating Language Structure and Use*. Cambridge University Press, New York, 1998.

[26] G. Blank and B. C. Reisdorf. The participatory web. *Information, Communication & Society*, 15(4):537–554, 2012.

[27] A. L. Blum and P. Langley. Selection of relevant features and examples in machine learning. *Artificial Intelligence*, 97(1):245–271, 1997.

[28] P. Bojanowski, E. Grave, A. Joulin, and T. Mikolov. Enriching word vectors with subword information. *Transactions of the Association for Computational Linguistics*, 5:135–146, 2017.

[29] W. M. Bolstad. *Introduction to Bayesian Statistics*. Wiley, Hoboken, 3rd edition, 2017.

[30] B. E. Boser, I. M. Guyon, and N. Vapnik, V. A training algorithm for optimal margin classifiers. In *Proceedings of the Fifth Annual Workshop on Computational Learning Theory*, COLT '92, pages 144–152, New York, NY, USA, 1992. ACM.

[31] S. Boyd and L. Vandenberghe. *Introduction to Applied Linear Algebra: Vectors, Matrices, and Least Squares*. Cambridge University Press, 2018.

[32] L. Breiman. Bagging predictors. *Machine Learning*, 24(2):123–140, 1996.

[33] L. Breiman. Random forests. *Machine Learning*, 45(1):5–31, 2001.

[34] L. Breiman. *Manual On Setting Up, Using, And Understanding Random Forests V3.1*, 2002. https://www.stat.berkeley.edu/~breiman/Using_random_forests_V3.1.pdf.

[35] L. Breiman. *Manual for Setting Up, Using, and Understanding Random Forest V4.0*, 2003. https://www.stat.berkeley.edu/~breiman/Using_random_forests_v4.0.pdf.

[36] L. Breiman, J. H. Friedman, R. A. Olshen, and C. J. Stone. *Classification and Regression Trees*. Wadsworth and Brooks, Monterey, CA, 1984.

[37] E. Brill. A simple rule-based part-of-speech tagger. In *Proceedings of ANLP-92, 3rd Conference on Applied Natural Language Processing*, pages 152–155, Trento, IT, 1992.

[38] R. D. Brown. Selecting and weighting n-grams to identify 1100 languages. In I. Habernal and V. Matoušek, editors, *Text, Speech, and Dialogue*, pages 475–483, Berlin, Heidelberg, 2013. Springer.

[39] T. Brychcín and M. Konopík. Hps: High precision stemmer. *Information Processing & Management*, 51(1):68–91, 2015.

[40] Q. Bsoul, J. Salim, and L. Q. Zakaria. An intelligent document clustering approach to detect crime patterns. *Procedia Technology*, 11:1181–1187, 2013.

[41] C. Buckley, J. Allan, and G. Salton. Automatic routing and retrieval using Smart: TREC-2. *Information Processing & Management*, 31(3):315–326, 1995.

[42] F. Can and E. A. Ozkarahan. Concepts and effectiveness of the cover-coefficient-based clustering methodology for text databases. *ACM Transactions on Database Systems*, 15(4):483–517, 1990.

[43] H. C. C. Carneiro, F. M. G. França, and P. M. V. Lima. Multilingual part-of-speech tagging with weightless neural networks. *Neural Networks*, 66:11–21, 2015.

[44] W. B. Cavnar and J. M. Trenkle. N-gram-based text categorization. In *Proceedings of the Third Annual Conference on Document Analysis and Information Retrieval (SDAIR)*, pages 161–175, 1994.

[45] M. Charrad, N. Ghazzali, V. Boiteau, and A. Niknafs. Nbclust: An r package for determining the relevant number of clusters in a data set. *Journal of Statistical Software*, 61(6), 2014.

[46] A. Chen and H. Ji. Graph-based clustering for computational linguistics: A survey. In *Proceedings of the 2010 Workshop on Graph-based Methods for Natural Language Processing, ACL 2010*, pages 1–9. Association for Computational Linguistic, 2010.

[47] J. Chen, H. Huang, S. Tian, and Y. Qu. Feature selection for text classification with Naïve bayes. *Expert Systems with Applications*, 36(3, Part 1):5432–5435, 2009.

[48] C. L. Chiang. *Statistical Methods of Analysis*. World Scientific, Singapore, 2003.

[49] E. Chisholm and T. G. Kolda. New term weighting formulas for the vector space method in information retrieval. Technical Report ORNL/TM-13756, Computer Science and Mathematics Division, Oak Ridge National Laboratory, Oak Ridge, 1999.

[50] F. Chollet and J. J. Allaire. *Deep Learning with R*. Manning Publications, 2018.

[51] J. Chung, K. Lee, R. Pedarsani, D. Papailiopoulos, and K. Ramchandran. Ubershuffle: Communication-efficient data shuffling for SGD via coding theory. In *31st Conference on Neural Information Processing Systems (NIPS 2017)*, 2017.

[52] K. W. Church and P. Hanks. Word association norms, mutual information, and lexicography. *Computational Linguistics*, 16(1):22–29, 1990.

[53] P. Cichosz. *Data Mining Algorithms: Explained Using R*. Wiley, Chichester, 2015.

[54] S. Clark. Statistical parsing. In A. Clark, C. Fox, and S. Lappin, editors, *The Handbook of Computational Linguistics and Natural Language Processing*, chapter 13, pages 333–363. Wiley-Blackwell, Chichester, 2013.

[55] B. Clarke, E. Fokoue, and H. H. Zhang. *Principles and Theory for Data Mining and Machine Learning*. Springer Series in Statistics. Springer, New York, 2009.

[56] S. Clinchant and F. Perronnin. Aggregating continuous word embeddings for information retrieval. In *Proceedings of the Workshop on Continuous Vector Space Models and their Compositionality*, pages 100–109. Association for Computational Linguistics, 2013.

[57] R. Collobert, J. Weston, L. Bottou, M. Karlen, K. Kavukcuoglu, and P. Kuksa. Natural language processing (almost) from scratch. *Journal of Machine Learning Research*, 12:2493–2537, 2011.

[58] R. Cotton. *Learning R*. O'Reilly, Sebastopol, CA, 2013.

[59] N. Cristianini and J. S. Taylor. *An Introduction to Support Vector Machines (and other kernal-base learning methods)*. Cambridge University Press, 2000.

[60] M. Culp, K. Johnson, and G. Michailidis. *The R Package Ada for Stochastic Boosting*, 2016. https://cran.r-project.org/web/packages/ada/ada.pdf.

[61] D. R. Cutting, D. R. Karger, J. O. Pedersen, and J. W. Tukey. Scatter/gather: A cluster-based approach to browsing large document collections. In *Proceedings of the 15th Annual International ACM SIGIR Conference on Research and Development in Information Retrieval*, SIGIR '92, pages 318–329, New York, NY, USA, 1992. ACM.

[62] S. Das and U. M. Cakmak. *Hands-On Automated Machine Learning: A beginner's guide to building automated machine learning systems using AutoML and Python.* Packt Publishing, 2018.

[63] M. Dash and H. Liu. Feature selection for clustering. In *Proceedings of the 4th Pacific-Asia Conference on Knowledge Discovery and Data Mining, Current Issues and New Applications*, PADKK '00, pages 110–121, London, UK, UK, 2000. Springer-Verlag.

[64] F. Dařena, J. Petrovský, J. Přichystal, and J. Žižka. Machine learning-based analysis of the association between online texts and stock price movements. *Inteligencia Artificial*, 21(61):95–110, 2018.

[65] F. Dařena and J. Žižka. Ensembles of classifiers for parallel categorization of large number of text documents expressing opinions. *Journal of Applied Economic Sciences*, 12(1):25–35, 2017.

[66] C. De Boom, S. Van Canneyt, T. Demeester, and B. Dhoedt. Representation learning for very short texts using weighted word embedding aggregation. *Pattern Recognition Letters*, 80(C):150–156, September 2016.

[67] M.-C. de Marneffe, T. Dozat, N. Silveira, K. Haverinen, F. Ginter, J. Nivre, and C. D. Manning. Universal stanford dependencies: A cross-linguistic typology. In *Proceedings of the Ninth International Conference on Language Resources and Evaluation (LREC-2014)*. European Language Resources Association (ELRA), 2014.

[68] C. M. De Vries, S. Geva, and A. Trotman. Document clustering evaluation: Divergence from a random baseline. In *Information Retrieval 2012 Workshop*, Dortmund, Germany, 2012. Technical University of Dortmund.

[69] S. Dey Sarkar, S. Goswami, A. Agarwal, and J. Aktar. A novel feature selection technique for text classification using naïve bayes. *International Scholarly Research Notices*, 2014.

[70] I. S. Dhillon and D. S. Modha. Concept decompositions for large sparse text data using clustering. *Machine Learning*, 42(1–2):143–175, 2001.

[71] T. Dietterich. Overfitting and undercomputing in machine learning. *Computing Surveys*, 27:326–327, 1995.

[72] T. Dietterich. An experimental comparison of three methods for constructing ensembles of decision trees: Bagging, boosting, and randomization. *Machine Learning*, 40(2):139–157, 2000.

[73] C. H. Q. Ding, X. He, H. Zha, M. Gu, and H. D. Simon. Spectral min-max cut for graph partitioning and data clustering. In *Proceedings 2001 IEEE International Conference on Data Mining*. IEEE, 2001.

[74] R. M. W. Dixon and A. Y. Aikhenvald. *Word: A Cross-linguistic Typology*. Cambridge University Press, 2002.

[75] G. Dong and H. Liu. *Feature Engineering for Machine Learning and Data Analytics*. Chapman & Hall/CRC Data Mining and Knowledge Discovery Series. CRC Press, Boca Raton, FL, 2018.

[76] R. O. Duda, P. E. Hart, and D. G. Stork. *Pattern Classification*. John Wiley & Sons, New York, 2007.

[77] S. T. Dumais. Improving the retrieval of information from external sources. *Behavior Research Methods, Instruments, & Computers*, 23(2):229–236, 1991.

[78] J. G. Dy. Unsupervised feature selection. In H. Liu and H. Motoda, editors, *Computational Methods of Feature Selection*, chapter 2, pages 19–39. Chapman & Hall/CRC, Boca Raton, 2008.

[79] Ecma International. The json data interchange format. Standard ECMA-404, October 2013.

[80] D. Eddelbuettel and R. Françcois. Rcpp: Seamless R and C++ integration. *Journal of Statistical Software*, 40(8), 2011.

[81] B. S. Everitt, S. Landau, and M. Leese. *Cluster Analysis*. Oxford University Press, New York, NY, 4th edition, 2001.

[82] S. Eyheramendy and D Madigan. A bayesian feature selection score based on naïve bayes models. In H. Liu and H. Motoda, editors, *Computational Methods of Feature Selection*, chapter 14, pages 277–294. Chapman & Hall/CRC, Boca Raton, 2008.

[83] D. Falbel. *R Interface to 'Keras'*, 2019. https://cran.r-project.org/web/packages/keras/keras.pdf.

[84] Y. C. Fang, S. Parthasarathy, and F. W. Schwartz. Using clustering to boost text classification. In *Workshop on Text Mining, TextDM 2001*, 2001.

[85] R. M. Fano. *Transmission of Information: A Statistical Theory of Communication*. The M.I.T. Press, Cambridge, MA, 1961.

[86] I. Feinerer, K. Hornik, and D. Meyer. Text mining infrastructure in R. *Journal of Statistical Software, Articles*, 25(5):1–54, 2008.

[87] R. Feldman and J. Sanger. *The Text Mining Handbook: Advanced Approaches in Analyzing Unstructured Data.* Cambridge: Cambridge University Press, 2007.

[88] G. Ferrano and L. Wanner. Labeling semantically motivated clusters of verbal relations. *Procesamiento del Lenguaje Natural*, 49:129–138, 2012.

[89] P. Flach. *Machine Learning: The Art and Science of Algorithms that Make Sense of Data.* Cambridge University Press, Cambridge, 2012.

[90] M. Flor. Four types of context for automatic spelling correction. *TAL*, 53(3):61–99, 2012.

[91] J. W. Foreman. *Data Smart: Using Data Science to Transform Information into Insight.* Wiley, Indianapolis, IN, 2014.

[92] G. Forman. Feature selection for text classification. In H. Liu and H. Motoda, editors, *Computational Methods of Feature Selection*, chapter 13, pages 257–276. Chapman & Hall/CRC, Boca Raton, 2008.

[93] C. Frankfort-Nachmias and A. Leon-Guerrero. *Social Statistics for a Diverse Society.* Pine Forge Press, Thousand Oaks, 5th edition, 2009.

[94] Y. Freund and R. E. Schapire. A decision-theoretic generalization of online learning and an application to boosting. *Journal of Computer and System Sciences*, 55:119–139, 1997.

[95] Y. Freund and R. E. Schapire. A short introduction to boosting. *Journal of Japanese Society for Artificial Intelligence*, 14(5):771–780, 1999.

[96] J. H. Friedman and U. Fayyad. On bias, variance, 0/1-loss, and the curse-of-dimensionality. *Data Mining and Knowledge Discovery*, 1:55–77, 1997.

[97] G. Gan, C. Ma, and J. Wu. *Data Clustering: Theory, Algorithms, and Applications.* ASA-SIAM Series on Statistics and Applied Probability. American Statistical Organization and Society for Industrial and Applied Mathematics, Philadelphia, 2007.

[98] A. Gelbukh and G. Sidorov. Zipf and Heaps' laws coefficients depend on language. In A. Gelbukh, editor, *CICLing 2001: Computational Linguistics and Intelligent Text Processing*, volume 2004 of *Lecture Notes in Computer Science*, pages 332–335. Springer, 2001.

[99] A. F. Gelbukh, M. Alexandrov, A. Bourek, and P. Makagonov. Selection of representative documents for clusters in a document collection. In *Proceedings of Natural Language Processing and Information Systems, 8th International Conference on Applications of Natural Language to Information Systems*, pages 120–126, 2003.

[100] A. Gelman, J. B. Carlin, J. S. Stern, D. B. Dunson, A. Vehtari, and D. B. Rubin. *Bayesian Data Analysis*. Texts in Statistical Science. Chapman & Hall/CRC, 3rd edition, 2013.

[101] Y. Goldberg. A primer on neural network models for natural language processing. *Journal of Artificial Intelligence Research*, 57(1):345–420, 2016.

[102] M. Goldszmidt, M. Najork, and S. Paparizos. Boot-strapping language identifiers for short colloquial postings. In H. Blockeel, K. Kersting, S. Nijssen, and F. Železný, editors, *Machine Learning and Knowledge Discovery in Databases*, pages 95–111, Berlin, Heidelberg, 2013. Springer Berlin Heidelberg.

[103] I. Goodfellow, Y. Bengio, and A. Courville. *Deep Learning*. Adaptive Computation and Machine Learning series. MIT Press, 2016.

[104] R. M. Gray. *Entropy and Information Theory*. Springer Science+Business Media, New York, 1990.

[105] J. K. Gross, J. Yellen, and M. Anderson. *Graph Theory and Its Applications*. Chapman and Hall/CRC, 2018.

[106] L. Grothe, E. W. De Luca, and A. Nürnberger. A comparative study on language identification methods. In *Proceedings of the International Conference on Language Resources and Evaluation, LREC 2008*, pages 980–985, 2008.

[107] Q. Gu, L. Zhu, and Z. Cai. Evaluation measures of the classification performance of imbalanced data sets. In Cai Z., Li Z., Kang Z., and Liu Y., editors, *Computational Intelligence and Intelligent Systems. ISICA 2009.*, volume 51, pages 461–471, Berlin, Heidelberg, 2009. Springer.

[108] Q. Guo and M. Zhang. Multi-documents automatic abstracting based on text clustering and semantic analysis. *Knowledge-Based Systems*, 22(6):482–485, 2009.

[109] N. Habash. *Introduction to Arabic Natural Language Processing*. Synthesis Lectures on Human Language Technologies. Morgan & Claypool Publishers, 2010.

[110] M. Halkidi, Y. Batistakis, , and M. Vazirgiannis. On clustering validation techniques. *Journal of Intelligent Information Systems*, 17(2–3):107–145, 2001.

[111] L. Hamel. *Knowledge Discovery with Support Vector Machines*. John Wiley & Sons, Inc., 2009.

[112] T. Hastie, R. Tibshirani, and J. Friedman. *The Elements of Statistical Learning: Data Mining, Inference, and Prediction*. Springer, 2nd edition, 2009.

[113] S. Haykin. *Neural Networks: A Comprehensive Foundation*. Prentice Hall, Upper Saddle River, NJ, 1994.

[114] H. S. Heaps. *Information Retrieval: Computational and Theoretical Aspects*. Academic Press, 1978.

[115] C. Hennig. What are the true clusters? *Pattern Recognition Letters*, 64:53–62, 2015.

[116] J. A. Hinojosa, M. Martín-Loeches, P. Casado, F. Muñoz, L. Carretié, C. Fernández-Frías, and M. A. Pozo. Semantic processing of open- and closed-class words: an event-related potentials study. *Cognitive Brain Research*, 11(3):397–407, 2001.

[117] K Hornik. R FAQ, 2017. https://CRAN.R-project.org/doc/FAQ/R-FAQ.html.

[118] J. Houvardas and E. Stamatatos. N-gram feature selection for authorship identification. In J. Euzenat and J. Domingue, editors, *Proceedings of the AIMSA conference*, volume 4183 of *LNAI*, pages 77–86, Berlin, 2006. Springer.

[119] H.-H. Hsu, C.-W. Hsieh, and M.-D. Lu. Hybrid feature selection by combining filters and wrappers. *Expert Systems with Applications*, 38(7):8144–8150, 2011.

[120] A. Huang. Similarity measures for text document clustering. In *Proceedings of the Sixth New Zealand Computer Science Research Student Conference*, pages 49–56, 2008.

[121] J. Z. Huang, J. Xu, M. Ng, and Ye. Y. Weighting method for feature selection in k-means. In H. Liu and H. Motoda, editors, *Computational Methods of Feature Selection*, chapter 10, pages 193–209. Chapman & Hall/CRC, Boca Raton, 2008.

[122] Z. Huang, M. K. Ng, and D. W.-L. Cheung. An empirical study on the visual cluster validation method with fastmap. In *Proceedings of the 7th international conference on database systems for advanced applications (DASFAA 2001)*, pages 84–91, Hong-Kong, 2001. Springer.

[123] T. Hudík. Machine translation within commercial companies. In J. Žižka and F. Dařena, editors, *Modern Computational Models of Semantic Discovery in Natural Language*, chapter 10, pages 256–272. IGI Global, Hershey, PA, 2015.

[124] A. S. Hussein. A plagiarism detection system for Arabic documents. *Advances in Intelligent Systems and Computing*, 323:541–552, 2015.

[125] R. Ihaka and R. Gentlman. R: A language for data analysis and graphics. *Journal of Computational and Graphical Statistics*, 5(3):299–314, 1996.

[126] H. Jackson and E. Z. Amvela. *Words, Meaning and Vocabulary: An Introduction to Modern English Lexicology*. Continuum, New York, 2000.

[127] A. K. Jain. Data clustering: 50 years beyond k-means. *Pattern Recognition Letters*, 31(8):651–666, 2010.

[128] A. K. Jain and R. C. Dubes. *Algorithms for Clustering Data*. Prentice Hall, Engelwood Cliffs, NJ, 1988.

[129] G. James, D. Witten, T. Hastie, and R. Tibshirani. *An Introduction to Statistical Learning: With Applications in R*. Springer, 2014.

[130] A. Janusz. Algorithms for similarity relation learning from high dimensional data. In J. F. Peters and A. Skowron, editors, *Transactions on Rough Sets XVII*, pages 174–292. Springer, Berlin, Heidelberg, 2014.

[131] T. Jauhiainen, K. Linden, and H. Jauhiainen. Evaluation of language identification methods using 285 languages. In *Proceedings of the 21st Nordic Conference of Computational Linguistics*, pages 183–191. Linkoping University Electronic Press, 2017.

[132] A. Jawlik. *Statistics from A to Z: Confusing Concepts Clarified*. Wiley, Hoboken, NJ, 2016.

[133] T. Jo. *Text Mining: Concepts, Implementation, and Big Data Challenge*, volume 45 of *Studies in Big Data*. Springer, 2019.

[134] T. Joachims. A statistical learning learning model of text classification for support vector machines. In *Proceedings of the 24th Annual International ACM SIGIR Conference on Research and Development in Information Retrieval*, SIGIR '01, pages 128–136, New York, NY, USA, 2001. ACM.

[135] T. Joachims. *Learning to Classify Text Using Support Vector Machines*. Kluwer Academic Publishers, Norwell, MA, 2002.

[136] B. Jongejan and H. Dalianis. Automatic training of lemmatization rules that handle morphological changes in pre-, in- and suffixes alike. In *Proceedings of the 47th Annual Meeting of the ACL and the 4th IJCNLP of the AFNLP*, pages 145–153. ACL and AFNLP, 2009.

[137] A. Jović, K. Brkić, and N. Bogunović. A review of feature selection methods with applications. In *38th International Convention on Information and Communication Technology, Electronics and Microelectronics (MIPRO)*, pages 1200–1205, May 2015.

[138] D. Jurafsky and J. H. Martin. *Speech and Language Processing: An Introduction to Natural Language Processing, Computational Linguistics, and Speech Recognition*. Prentice Hall, 2009.

[139] I. Kanaris, K. Kanaris, I. Houvardas, and E. Stamatatos. Words versus character n-grams for anti-spam filtering. *International Journal on Artificial Intelligence Tools*, 16(6):1047–1067, 2007.

[140] N. N. Karanikolas. Supervised learning for building stemmers. *Journal of Information Science*, pages 1–14, 2015.

[141] G. Karypis. Cluto: A Clustering Toolkit. Technical Report 02-017, University of Minnesota, Department of Computer Science, 2003.

[142] L. Kaufman and P. J. Rousseeuw. *Finding Groups in Data: An Introduction to Cluster Analysis*. Wiley, Hoboken, NJ, 2005.

[143] B. Keith, E. Fuentes, and C. Meneses. A hybrid approach for sentiment analysis applied to paper reviews. In *Proceedings of ACM SIGKDD Conference*, 2017.

[144] J. Kent Martin and D. S. Hirschberg. The time complexity of decision tree induction. Technical Report 95–27, Department of Information and Computer Science. University of California, Irvine, CA, 92717, August 1995.

[145] E. Keogh and A. Mueen. Curse of dimensionality. In C. Sammut and G. I. Webb, editors, *Encyclopedia of Machine Learning and Data Mining*, pages 314–315. Springer US, Boston, MA, 2017.

[146] R. Khoury, F. Karray, and M. S. Kamel. Keyword extraction rules based on a part-of-speech hierarchy. *International Journal of Advanced Media and Communication*, 2(2):138–153, 2008.

[147] M. Kleppmann. *Designing Data-Intensive Applications: The Big Ideas Behind Reliable, Scalable, and Maintainable Systems.* O'Reilly Media, Sebastopol, CA, 2017.

[148] T. Kocmi and B. Bojar. An Exploration of Word Embedding Initialization in Deep-Learning Tasks. *CoRR*, abs/1711.09160, 2017.

[149] J. K. Korpela. *Unicode Explained.* O'Reilly, Beijing, 2006.

[150] M. Kubat. *An Introduction to Machine Learning.* Springer, 2015.

[151] M. Kubát. *An Introduction to Machine Learning.* Springer International Publishing, 2nd edition, 2017.

[152] S. Kudyba. *Big Data, Mining, and Analytics: Components of Strategic Decision Making.* CRC Press, Boca Raton, FL, 2014.

[153] M. Kuhn. *Classification and Regression Training*, 2019. https://cran.r-project.org/web/packages/caret/caret.pdf.

[154] M. Kuhn, S. Weston, M. Culp, N. Coulter, and R. Quinlan. *C5.0 Decision Trees and Rule-Based Models*, 2018. https://cran.r-project.org/web/packages/C50/C50.pdf.

[155] K. Kukich. Techniques for automatically correcting words in text. *ACM Computing Surveys*, 24(4):377–439, 1992.

[156] B. Kulis. Metric learning: A survey. *Machine Learning*, 5(4):287–364, 2012.

[157] M. Kusner, Y. Sun, N. Kolkin, and K. Weinberger. From word embeddings to document distances. In F. Bach and B. Blei, editors, *Proceedings of the 32nd International Conference on Machine Learning*, volume 37 of *Proceedings of Machine Learning Research*, page 957966, Lille, France, 07–09 Jul 2015. PMLR.

[158] A. Kyriakopoulou and T. Kalamboukis. Text classification using clustering. In *In Proceedings of the ECML-PKDD Discovery Challenge Workshop*, 2006.

[159] M. Labani, P. Moradi, F. Ahmadizar, and M. Jalili. A novel multivariate filter method for feature selection in text classification problems. *Engineering Applications of Artificial Intelligence*, 70:25–37, 2018.

[160] B. Lambert. *A Student's Guide to Bayesian Statistics.* SAGE Publications, 2018.

[161] P. Larranaga, D. Atienza, J. Diaz-Rozo, A. Ogbechie, C. E. Puerto-Santana, and C. Bielza. *Industrial Applications of Machine Learning*. Chapman & Hall/CRC Data Mining and Knowledge Discovery Series. CRC Press, Boca Raton, FL, 2018.

[162] Q. Le and T. Mikolov. Distributed representations of sentences and documents. In *Proceedings of the 31st International Conference on International Conference on Machine Learning – Volume 32*, ICML'14, pages II–1188–II–1196. JMLR.org, 2014.

[163] R. Lebret and R. Collobert. N-gram-based low-dimensional representation for document classification. In *ICLR 2015 Workshop Track*, 2105.

[164] A. Lesne. Shannon entropy: a rigorous notion at the crossroads between probability, information theory, dynamical systems and statistical physics. *Mathematical Structures in Computer Science*, 24(3), 2014.

[165] O. Levy and Y. Goldberg. Neural word embedding as implicit matrix factorization. In Z. Ghahramani, M. Welling, C. Cortes, N. D. Lawrence, and K. Q. Weinberger, editors, *Advances in Neural Information Processing Systems 27*, pages 2177–2185. Curran Associates, Inc., 2014.

[166] O. Levy, Y. Goldberg, and I. Dagan. Improving distributional similarity with lessons learned from word embeddings. *Transactions of the Association for Computational Linguistics*, 3:211–225, 2015.

[167] D. D. Lewis, Y. Yang, T. G. Rose, and F. Li. Rcv1: A new benchmark collection for text categorization research. *Journal of Machine Learning Research*, 5:361–397, 2004.

[168] C. Li, A. Sun, J. Weng, and Q. He. Tweet segmentation and its application to named entity recognition. *IEEE Transactions on Knowledge and Data Engineering*, 27(2):558–570, 2015.

[169] S. Li, R. Xia, C. Zong, and Huang. C.-R. A framework of feature selection methods for text categorization. In *Proceedings of the 47th Annual Meeting of the ACL and the 4th IJCNLP of the AFNLP*, pages 692–700, Suntec, Singapore, 2–7 August 2009. ACL and AFNLP.

[170] X. Li and B. Liu. Learning to classify texts using positive and unlabeled data. In *Proceedings of the 18th International Joint Conference on Artificial Intelligence*, IJCAI'03, pages 587–592, San Francisco, CA, USA, 2003. Morgan Kaufmann Publishers Inc.

[171] A. Liaw and M. Wiener. *Breiman and Cutler's Random Forests for Classification and Regression*, 2018. https://cran.r-project.org/web/packages/randomForest/randomForest.pdf.

[172] Y. Lin, J.-B. Michel, E. L. Aiden, J. Orwant, W. Brockman, and S. Petrov. Syntactic annotations for the Google books ngram corpus. In *Proceedings of the ACL 2012 System Demonstrations*, ACL '12, pages 169–174, Stroudsburg, PA, USA, 2012. Association for Computational Linguistics.

[173] T. W. Ling, M. L. Lee, and G. Dobbie. *Semistructured Database Design*. Springer, 2005.

[174] H. Liu and H. Motoda. *Feature Selection for Knowledge Discovery and Data Mining*. Kluwer Academic Publishers, Norwell, MA, USA, 1998.

[175] H. Liu and H. Motoda. Less is more. In H. Liu and H. Motoda, editors, *Computational Methods of Feature Selection*, chapter 1, pages 3–18. Chapman & Hall/CRC, Boca Raton, 2008.

[176] H. Liu, J. Sun, L. Liu, and H. Zhang. Feature selection with dynamic mutual information. *Pattern Recognition*, 42(7):1330–1339, 2009.

[177] K. Liu, A. Bellet, and F. Sha. Similarity learning for high-dimensional sparse data. In *Proceedings of the 18th International Conference on Artificial Intelligence and Statistics (AISTATS) 2015*, volume 38 of *JMLR Workshop and Conference Proceedings*, pages 653–662. JMLR.org, 2015.

[178] L. Liu, J. Kang, J. Yu, and Z. Wang. A comparative study on unsupervised feature selection methods for text clustering. In *2005 International Conference on Natural Language Processing and Knowledge Engineering*, pages 597–601. IEEE, 2005.

[179] T. Liu, S. Liu, Z. Chen, and W-Y. Ma. An evaluation on feature selection for text clustering. In *Proceedings of the Twentieth International Conference on Machine Learning (ICML-2003)*, Washington DC, 2003.

[180] Y. Liu, J. Mostafa, and W. Ke. A fast online clustering algorithm for scatter/gather browsing. Technical Report TR-2007-06, NC School of Information and Library Science, Chapel Hill, NC, 2007.

[181] K. E. Lochbaum and L. A. Streeter. Comparing and combining the effectiveness of latent semantic indexing and the ordinary vector space model for information retrieval. *Information Processing & Management*, 25(6):665–676, 1989.

[182] M. Majka. *High Performance Implementation of the Naive Bayes Algorithm.* 2019. https://cran.r project.org/web/packages/naivebayes/naivebayes.pdf.

[183] N. Maltoff. *The Art of R Programming: A Tour of Statistical Software Design*. No Starch Press, San Francisco, CA, 2011.

[184] B. Mandelbrot. Structure formelle des textes et communication. *Word*, 10:1–27, 1954.

[185] C. D. Manning, P. Raghavan, and H. Schütze. *Introduction to Information Retrieval*. Cambridge University Press, 2008.

[186] C. D. Manning and H. Schütze. *Foundations of Statistical Natural Language Processing*. The MIT Press, Cambridge, Massachusetts, 1999.

[187] S. Marsland. *Machine Learning: An Algorithmic Perspective*. Chapman and Hall/CRC, Boca Raton, FL, 2nd edition, 2014.

[188] E. Mayfield and C. Penstein-Rosé. Using feature construction to avoid large feature spaces in text classification. In *Proceedings of the 12th Annual Conference on Genetic and Evolutionary Computation*, GECCO '10, pages 1299–1306, New York, NY, USA, 2010. ACM.

[189] D. Maynard, K. Bontcheva, and I. Augenstein. *Natural Language Processing for the Semantic Web*. Morgan & Claypool, 2017.

[190] J. McAuley. Amazon product data. http://jmcauley.ucsd.edu/data/amazon/. Accessed on June 8, 2018.

[191] R. McElreath. *Statistical Rethinking: A Bayesian Course with Examples in R and Stan*. CRC Press, 2016.

[192] J. Mena. *Investigative Data Mining for Security and Criminal Detection*. Butterworth-Heinemann, Newton, MA, USA, 2002.

[193] D. Meyer, E. Dimitriadou, K. Hornik, F. Weingessel, A. Leisch, C.-C. Chang, and C.-C. Lin. *Misc Functions of the Department of Statistics, Probability Theory Group (Formerly: E1071), TU Wien*, 2019. https://cran.r-project.org/web/packages/e1071/e1071.pdf.

[194] T. Mikolov, I. Sutskever, K. Chen, G. S. Corrado, and J. Dean. Distributed representations of words and phrases and their compositionality. In C. J. C. Burges, L. Bottou, M. Welling, Z. Ghahramani, and K. Q. Weinberger, editors, *Advances in Neural Information Processing Systems 26*, pages 3111–3119. Curran Associates, Inc., 2013.

[195] T. Mikolov, W. Yih, and G. Zweig. Linguistic regularities in continuous space word representations. In *Proceedings of NAACL-HLT 2013*, pages 746–751. Association for Computational Linguistic, 2013.

[196] G. Miner, J. Elder, T. Hill, R. Nisbet, D. Delen, and A. Fast. *Practical Text Mining and Statistical Analysis for Non-structured Text Data Applications*. Academic Press, Inc., Orlando, FL, USA, 1st edition, 2012.

[197] B. Mirkin. *Mathematical Classification and Clustering.* Kluwer Academic Publishers, Dodrecht, 1996.

[198] B. Mirkin. *Clustering: A Data Recovery Approach.* CRC Press, Boca Raton, FL, 2nd edition, 2013.

[199] T. M. Mitchell. *Machine Learning.* WCB/McGraw-Hill, 1997.

[200] D. Mladenić and M. Grobelnik. Feature selection for unbalanced class distribution and naive bayes. In *In Proceedings of the 16th International Conference on Machine Learning (ICML,* pages 258–267. Morgan Kaufmann Publishers, 1999.

[201] Dunja Mladenić. Feature selection in text mining. In C. Sammut and Geoffrey I. Webb, editors, *Encyclopedia of Machine Learning and Data Mining,* pages 1–5. Springer US, Boston, MA, 2016.

[202] A. Mountassir, H. Benbrahim, and I. Berrada. Addressing the problem of unbalanced data sets in sentiment analysis. In *Proceedings of the International Conference on Knowledge Discovery and Information Retrieval (KDIR-2012),* pages 306–311, 2012.

[203] S. Munzert, C. Rubba, P. Meiner, and D. Nyhuis. *Automated Data Collection with R: A Practical Guide to Web Scraping and Text Mining.* Wiley, Chichester, 1st edition, 2014.

[204] P. Murrell. *R Graphics.* CRC Press, Boca Raton, 2nd edition, 2011.

[205] D. Nadeau and S. Sekine. A survey of named entity recognition and classification. *Lingvisticae Investigationes,* 30(1):3–26, 2007.

[206] M.-J. Nederhof and G. Satta. Theory of parsing. In A. Clark, C. Fox, and S. Lappin, editors, *The Handbook of Computational Linguistics and Natural Language Processing,* chapter 4, pages 105–130. Wiley-Blackwell, Chichester, 2013.

[207] J. Nothman, H. Qin, and R. Yurchak. Stop word lists in free open-source software packages. In *Proceedings of Workshop for NLP Open Source Software (NLP-OSS),* pages 7–12. Association for Computational Linguistics, 2018.

[208] Y. Ohsawa and K. Yada. *Data Mining for Design and Marketing.* Chapman and Hall/CRC, Boca Raton, FL, 2017.

[209] C. Ozgur, T. Colliau, G. Rogers, Z. Hughes, and E. Myer-Tyson. Matlab vs. Python vs. R. *Journal of Data Science,* 15:355–372, 2017.

[210] D. D. Palmer. Tokenisation and sentence segmentation. In R. Dale, H. Moisl, and H. Somers, editors, *A Handbook of Natural Language Processing*, pages 11–36. Dekker, New York, 2000.

[211] D. D. Palmer. Text preprocessing. In *Handbook of Natural Language Processing*, pages 9–30. Chapman & Hall/CRC, 2nd edition, 2010.

[212] M. Pelillo, editor. *Similarity-Based Pattern Analysis and Recognition*. Advances in Computer Vision and Pattern Recognition. Springer Verlag London, 2013.

[213] R. D. Peng. *R Programming for Data Science*. lulu.com, 2016.

[214] J. Pennington, R. Socher, and C. D. Manning. Glove: Global vectors for word representation. In *EMNLP*, volume 14, pages 1532–1543, 2014.

[215] G. Pethö and E. Mózes. An n-gram-based language identification algorithm for variable-length and variable-language texts. *Argumentum*, 10:56–82, 2014.

[216] N. Polettini. The Vector Space Model in Information Retrieval – Term Weighting Problem, 2004.

[217] M. F. Porter. An algorithm for suffix stripping. *Program*, 14(3):130–137, 1980.

[218] M. F. Porter. Snowball: A language for stemming algorithms, October 2001. http://snowball.tartarus.org/texts/introduction.html.

[219] M. Prigmore. *An Introduction to Databases with Web Applications*. Pearson Education, Harlow, 2008.

[220] J. R. Quinlan. *C4.5: Programs for Machine Learning*. Morgan Kaufmann, 1993.

[221] B. Ratner. *Statistical and Machine-Learning Data Mining: Techniques for Better Predictive Modeling and Analysis of Big Data*. CRC Press, Boca Raton, FL, 2012.

[222] C. K. Reddy and B. Vinzamuri. A survey of partitional and hierarchical clustering algorithms. In C. C. Aggarwal and C. K. Reddy, editors, *Data Clustering: Algorithms and Applications*, pages 87–110. CRC Press, Boca Raton, FL, 2014.

[223] D. K. S. Reddy, S. K. Dash, and A. K. Pujari. New malicious code detection using variable length n-grams. In *Proceedings of the Second International Conference on Information Systems Security*, ICISS'06, pages 276–288, Berlin, Heidelberg, 2006. Springer.

[224] RuleQuest Research. Free software downloads, 2018. https://rulequest.com/download.html.

[225] J. C. Reynar and A. Ratnaparkhi. A maximum entropy approach to identifying sentence boundaries. In *Proceedings of the Fifth Conference on Applied Natural Language Processing*, ANLC '97, pages 16–19, Stroudsburg, PA, USA, 1997. Association for Computational Linguistics.

[226] B. Ripley and W. Venables. *Functions for Classification*, 2019. https://cran.r-project.org/web/packages/class/class.pdf.

[227] S. Robertson. Understanding inverse document frequency: On theoretical arguments for idf. *Journal of Documentation*, 60(5):503–520, 2004.

[228] Y. Roh, G. Heo, and S. E. Whang. A survey on data collection for machine learning: a big data - ai integration perspective. *CoRR*, abs/1811.03402, 2018.

[229] L. Rokach. A survey of clustering algorithms. In O. Maimon and L. Rokach, editors, *Data Mining and Knowledge Discovery Handbook*, pages 269–298. Springer, New York, 2nd edition, 2010.

[230] H. C. Romesburg. *Cluster Analysis for Researchers*. Lulu Press, Raleigh, NC, 2004.

[231] P. J. Rousseeuw. Silhouettes: A graphical aid to the interpretation and validation of cluster analysis. *Journal of Computational and Applied Mathematics*, 20:53–65, 1987.

[232] M. Roux. A comparative study of divisive and agglomerative hierarchical clustering algorithms. *Journal of Classification*, 2018.

[233] D. Salomon. *Variable-length Codes for Data Compression*. Springer, London, 2007.

[234] G. Salton. Cluster search strategies and the optimization of retrieval effectiveness. In G. Salton, editor, *The SMART Retrieval System – Experiments in Automatic Document Processing*, pages 223–242. Prentice-Hall, Upper Saddle River, NJ, 1971.

[235] G. Salton and C. Buckley. Term-weighting approaches in automatic text retrieval. *Information Processing & Management*, 24(5):513–523, 1988.

[236] McGill M. J. Salton G. *Introduction to Modern Information Retrieval*. McGraw Hill, New York, 1983.

[237] G. Sanchez. Handling and processing strings in r, 2013. http://www.gastonsanchez.com/Handling_and_Processing_Strings_in_R.pdf.

[238] K. R. Saoub. *A Tour through Graph Theory*. Chapman and Hall/CRC, 2017.

[239] J. Saratlija, J. Šnajder, and B. Dalbelo Bašić. Unsupervised topic-oriented keyphrase extraction and its application to croatian. In I. Habernal and V. Matoušek, editors, *14th International Conference on Text, Speech and Dialogue*, volume 6836 of *Lecture Notes in Artificial Intelligence*, pages 340–347, 2011.

[240] S. Savaresi, D. Boley, S. Bittanti, and G. Gazzaniga. Choosing the cluster to split in bisecting divisive clustering algorithms. In R. Grossman, J. Han, V. Kumar, H. Mannila, and R. Motwani, editors, *Second SIAM International Conference on Data Mining (SDM'2002)*, Arlington, VA, 2002.

[241] P. Schachter and T. Shopen. Parts-of-speech systems. In T. Shopen, editor, *Language Typology and Syntactic Description*, volume 1, pages 1–60. Cambridge University Press, Cambridge, 2nd edition, 2007.

[242] S. E. Schaeffer. Graph clustering. *Computer science review*, 1:27–64, 2007.

[243] B. Schölkopf. *Learning with Kernels: Support Vector Machines, Regularization, Optimization, and Beyond*. Adaptive Computation and Machine Learning series. The MIT Press, 2001.

[244] S. Schulte im Walde. Experiments on the automatic induction of german semantic verb classes. *AIMS: Arbeitspapiere des Instituts für Maschinelle Sprachverarbeitung*, 9(2), 2003.

[245] F. Sebastiani. Machine learning in automated text categorization. *ACM Computing Surveys*, 1:1–47, 2002.

[246] G. A. F. Seber. *Multivariate Observations*. John Wiley & Sons, Hoboken, 1984.

[247] M. Seeger. Learning with labeled and unlabeled data. Technical report, University of Edinburgh, 2001.

[248] Y. Shafranovich. Common format and mime type for comma-separated values (csv) files. Technical Report RFC 4180, Network Working Group, October 2005.

[249] C. E. Shannon. A mathematical theory of communication. *The Bell System Technical Journal*, 27(3):379–423, 1948.

[250] G. Shmueli, N. R. Patel, and Bruse P. C. *Data Mining for Business Intelligence: Concepts, Techniques, and Applications in Microsoft Office Excel with XLMiner*. Wiley, Hoboken, NJ, 2010.

[251] R. L. Sims. *Bivariate Data Analysis: A Practical Guide*. Nova Science Publishers, Hauppauge, NY, 2014.

[252] A. K. Singhal. *Term Weighting Revisited*. PhD thesis, Faculty of the Graduate School of Cornell University, 1997.

[253] R. Socher, A. Perelygin, J. Wu, J. Chuang, C. D. Manning, . Ng, and C. Potts. Recursive deep models for semantic compositionality over a sentiment treebank. In *Proceedings of the 2013 Conference on Empirical Methods in Natural Language Processing*, pages 1631–1642. Association for Computational Linguistics, 2013.

[254] A. Srivastava and M. Sahami. *Text Mining: Classification, Clustering, and Applications*. CRC Press, Boca Raton, 2009.

[255] B. Stein and S. M. zu Eissen. Automatic document categorization. In A. Günter, R. Kruse, and B. Neumann, editors, *KI 2003: Advances in Artificial Intelligence*, pages 254–266, Berlin, Heidelberg, 2003. Springer.

[256] M. Steinbach, G. Karypis, and V. Kumar. A comparison of document clustering techniques. In *KDD Workshop on Text Mining*, 2000.

[257] B. Steinley. K-means clustering: a half-century synthesis. *The British Journal of Mathematical and Statistical Psychology*, 59 Pt 1:1–34, 2006.

[258] D. J Stracuzzi. Randomized feature selection. In H. Liu and H. Motoda, editors, *Computational Methods of Feature Selection*, chapter 3, pages 41–62. Chapman & Hall/CRC, Boca Raton, 2008.

[259] A. Taylor, M. Marcus, and B. Santorini. The Penn Treebank: An overview. In A. Abeillé, editor, *Treebanks: Building and Using Parsed Corpora*, pages 5–22. Springer Netherlands, Dordrecht, 2003.

[260] P. Teetor. *R Cookbook: Proven Recipes for Data Analysis, Statistics, and Graphics*. O'Reilly, Sebastopol, CA, 2011.

[261] L. Torgo. *Data Mining with R: Learning with Case Studies*. Chapman and Hall/CRC, Boca Raton, FL, 2nd edition, 2017.

[262] Y.-H. Tseng, C.-J. Lin, and Y.-I Lin. Text mining techniques for patent analysis. *Information Processing & Management*, 43(5):1216–1247, 2007.

[263] P. D. Turney and P. Pantel. From frequency to meaning: Vector space models of semantics. *Journal of Artificial Intelligence Research*, 37(1):141–188, 2010.

[264] T. C. Urdan. *Statistics in Plain English*. Lawrence Erlbaum Associates, Mahwah, NJ, 2005.

[265] J. Vasagar. Germany drops its longest word: Rindfleischeti... *The Telegraph*, Jun 2013. https://www.telegraph.co.uk/news/worldnews/europe/germany/10095976/ Germany-drops-its-longest-word-Rindfleischeti....html.

[266] S. V. Vaseghi. *Advanced Digital Signal Processing and Noise Reduction*. John Wiley & Sons, Ltd, Chichester, 2006.

[267] N. X. Vinh, J. Epps, and J. Bailey. Information theoretic measures for clusterings comparison: Variants, properties, normalization and correction for chance. *Journal of Machine Learning Research*, 11:2837–2854, 2010.

[268] U. von Luxburg, R. C. Williamson, and I. Guyon. Clustering: Science or art? In I. Guyon, G. Dror, V. Lemaire, G. Taylor, and D. Silver, editors, *Proceedings of ICML Workshop on Unsupervised and Transfer Learning*, volume 27 of *Proceedings of Machine Learning Research*, pages 65–79, Bellevue, Washington, USA, 02 Jul 2012. PMLR.

[269] E. Vylomova, L. Rimell, T. Cohn, and T. Baldwin. Take and took, gaggle and goose, book and read: Evaluating the utility of vector differences for lexical relation learning. In *Proceedings of the 54th Annual Meeting of the Association for Computational Linguistics (Volume 1: Long Papers)*, pages 1671–1682. Association for Computational Linguistics, 2016.

[270] J. Žižka, A. Svoboda, and F. Dařena. Selecting characteristic patterns of text contributions to social networks using instance-based learning algorithm ibl-2. In S. Kapounek and V. Krøutilová, editors, *Enterprise and Competitive Environment*, pages 971–980, Brno, 2017. Mendel University in Brno, Faculty of Business and Economics.

[271] H. Wachsmuth. *Text Analysis Pipelines: Towards Ad-hoc Large-Scale Text Mining*, volume 9383 of *Lecture Notes in Computer Science*. Springer, 2015.

[272] K. Wagstaff, C. Cardie, S. Rogers, and S. Schroedl. Constrained k-means clustering with background knowledge. In *ICML '01 Proceedings of the Eighteenth International Conference on Machine Learning*, pages 577–584. Morgan Kaufmann, 2001.

[273] S. Wang and Manning. C. D. Baselines and bigrams: Simple, good sentiment and topic classification. In *Proceedings of the 50th Annual Meeting of the Association for Computational Linguistics*, pages 90–94. Association for Computational Linguistic, 2012.

[274] Z. Wang and X. Xue. Multi-class support vector machine. In Y. Ma and G. Guo, editors, *Support Vector Machines Applications*, chapter 2, pages 23–48. Springer International Publishing, Cham, 2014.

[275] S. M. Weiss, N. Indurkhya, T. Zhang, and F. J. Damerau. *Text Mining: Predictive Methods for Analyzing Unstructured Information.* Springer New York, New York, NY, 2005.

[276] L. Weng. Learning word embedding, 2017. https://lilianweng.github.io/lil-log/2017/10/15/learning-word-embedding.html. Accessed 2019-02-14.

[277] J. J. Whang, I. S. Dhillon, and D. F. Gleich. *Non-exhaustive, Overlapping K-means*, pages 936–944. Society for Industrial and Applied Mathematics Publications, 2015.

[278] W. J. Wilbur and K. Sirotkin. The automatic identification of stop words. *Journal of Information Science*, 18(1):45–55, February 1992.

[279] I. H. Witten. Text mining. In *The Practical Handbook of Internet Computing.* Chapman and Hall/CRC, 2004.

[280] I. H. Witten, E. Frank, and M. A. Hall. *Data Mining: Practical Machine Learning Tools and Techniques.* Morgan Kaufmann, San Francisco, 2011.

[281] D. H. Wolpert. The lack of a priori distinctions between learning algorithms. *Neural Computation*, 8(7):1341–1390, 1996.

[282] D. H. Wolpert. The supervised learning no-free-lunch theorems. In *Proceedings of the 6th Online World Conference on Soft Computing in Industrial Applications*, pages 25–42, 2001.

[283] E. P. Xing, A. Y. Ng, M. I. Jordan, and S. Russell. Distance metric learning, with application to clustering with side-information. In *Proceedings of the 15th International Conference on Neural Information Processing Systems*, pages 521–528, 2002.

[284] G. Xu, Y. Zong, and Z. Yang. *Applied Data Mining.* CRC Press, Boca Raton, FL, 2013.

[285] R. Xu and D. C. Wunsch. *Clustering.* Wiley, Hoboken, NJ, 2009.

[286] Y. Yang and J. O. Pedersen. A comparative study on feature selection in text categorization. In *Proceedings of the Fourteenth International Conference on Machine Learning (ICML'97)*, pages 412–420, San Francisco, USA, 1997. Morgan Kaufmann Publishers.

[287] A. Zell, N. Mache, R. Hübner, G. Mamier, M. Vogt, M. Schmalzl, and K.-U. Herrmann. Snns (Stuttgart neural network simulator). In J. Skrzypek, editor, *Neural Network Simulation Environments*, pages 165–186. Springer US, Boston, MA, 1994.

[288] O. Zennaki, N. Semmar, and L. Besacier. Unsupervised and lightly supervised part-of-speech tagging using recurrent neural networks. In *29th Pacific Asia Conference on Language, Information and Computation (PACLIC)*, pages 133–142, Shangai, China, 2015.

[289] H. Zha, X. He, C. H. Q. Ding, M. Gu, and H. D. Simon. Bipartite graph partitioning and data clustering. In *Proceedings of the 2001 ACM CIKM International Conference on Information and Knowledge Management*, pages 25–32. ACM, 2001.

[290] Q. Zhang, J. Kang, J. Qian, and X. Huang. Continuous word embeddings for detecting local text reuses at the semantic level. In *Proceedings of the 37th International ACM SIGIR Conference on Research &#38; Development in Information Retrieval*, SIGIR '14, pages 797–806, New York, NY, USA, 2014. ACM.

[291] Y. Zhao and G. Karypis. Criterion functions for document clustering: Experiments and analysis. Technical Report 01-40, University of Minnesota, Department of Computer Science, 2001.

[292] Y. Zhao and G. Karypis. Comparison of agglomerative and partitional document clustering algorithms. In I. S. Dhillon and J. Kogan, editors, *Proceedings of the Workshop on Clustering High Dimensional Data and Its Applications at the Second SIAM International Conference on Data Mining*, pages 83–93, Philadelphia, 2002. SIAM.

[293] Y. Zhao and G. Karypis. Empirical and theoretical comparisons of selected criterion functions for document clustering. *Machine Learning*, 55(3):311–331, 2004.

[294] Y. Zhao and G. Karypis. Hierarchical clustering algorithms for document datasets. *Data Mining and Knowledge Discovery*, 10(2):141–168, 2005.

[295] J. Žižka and F. Dařena. Mining significant words from customer opinions written in different natural languages. In I. Habernal and V. Matoušek, editors, *Text, Speech and Dialogue*, pages 211–218, Berlin, Heidelberg, 2011. Springer.

# Index